THE EXPERTS GUIDE TO BUSINESS TELEPHONE SYSTEMS

THE EXPERTS GUIDE TO BUSINESS TELEPHONE SYSTEMS

Andrew L. Taub

CONTENTS

INTRODUCTION

Business Telephone Systems

The business world today is fast-paced and highly competitive; you need to be able to keep up with the latest telephony technology.

Business Telephone Systems help you stay in touch with your employees and customers, becoming the very foundation of your company.

The businesses that recognize the importance of a quality business telephone system are able to find powerful solutions that give their business the edge in an ever-changing market.

Whether you are a call-center, a restaurant or a national company with multiple locations there is a business telephone system available to suit your needs. The problem of course is navigating the complex and technical landscape involved in buying your ideal business telephone system.

Many business owners do not have any idea where to begin, so they rely on product advertising and sales hype that telephone system manufacturers release into the world. This can result in your company making the wrong decision about the telephone system you really need.

Efficiency and *reliability* must be the two pillars that drive your pursuit for the best possible business telephone system available for your company.

You simply cannot rely on empty sales talk to make these crucial communication decisions.

If your business does not communicate efficiently, it will not expand as rapidly as possible. There are numerous and continuous advances in technology happening all the time in the communications sector—and you need to be aware of these before selecting and installing your business telephone system.

This book is about familiarizing you with technological advancements in the communications field and it is about teaching you which communication systems and configurations are available to you in the market. There is a large variety of choice out there.

Once you have read the information here, you will be fully equipped to make an educated decision about your company's business telephone system. In the long run, this will mean more functionality, greater control and vastly improved communication capabilities.

Together, these advancements will allow your company to operate on a higher level, as you use the best business telephone system to help your business take on new markets, find new customers, improve operational efficiency and grow consistently.

Welcome to the one book on business telephone systems that will answer all of your questions. Use it as a reference guide and educate yourself on what is possible in this intriguing communications niche.

CHAPTER 1

Business Telephone Systems 101

"Precision of communication is important, more important than ever, in our era of hair trigger balances, when a false or misunderstood word may create as much disaster as a sudden thoughtless act."

– JAMES THURBER

For many small to medium-sized businesses, choosing the appropriate business telephone system can be incredibly challenging. This is because there are dozens of options available, all of which are fairly technical in nature.

If your company needs to have a new telephone system installed, then this information is invaluable to the future development of your company. Not only will you be able to select the right option, you will also be better equipped to maximize practical use of your system once selected. This is business telephone systems 101.

WHAT ARE BUSINESS TELEPHONE SYSTEMS?

The right business telephone system for your small to medium- sized company will help you improve productivity; it enhances collaboration and saves you money in the long run.

A business telephone system can be defined as the collection of multiple telephones that are all interconnected, allowing the business to utilize a variety of features.

These features can include intercom, conference calling, cellphone twinning, find me follow me, call transferring, off premise extension, call forwarding, call screening, and the use of automated attendant and voice messaging just to name a few.

Today, most telephone systems are scalable. In other words, as a business's communication needs grow; the telephone system easily migrates to the larger line capacity. Whether your system consists of a few telephones, hundreds or even thousands of telephones; they can be expanded to meet your communications needs both now and in the future.

Telephone systems utilize the traditional public switched telephone network (PSTN) or the Internet. These days it is becoming fairly commonplace to use both, for greater functionality and backup.

In order to make an educated decision which business telephone system will work best for your company, you need to gain a greater understanding of the roles these systems play in a business context.

The telephone system that you choose should strengthen the communication capabilities of your company, increase productivity and streamline business processes.

UNDERSTANDING WHAT YOU NEED

The first thing you must understand about business telephone systems is that it caters to any variety of needs that your company may have. There are large selections of benefits, features and configurations that you will have to work through.

The initial step of course, is about understanding what your business needs.

The right phone system will equip your employees with the tools they need to conduct business efficiently. This begs the question —what tools do they really need?

- Unified Communications
- Mobility Solutions
- Remote or Home Worker solutions
- Video Conferencing
- Appointment Reminder Applications
- Call Center Functionality
- Emergency Response Notification
- Rich Presence Technology
- Unified Messaging
- Wireless Solutions
- Call Recording

These are just some of the many features that you have to think about, before trying to select your business telephone system. A brand new system is going to rapidly change your company. It will be an adjustment, as your employees will be able to achieve higher level communications and customer service.

Integrating the right business telephone system into your business, changes the way you conduct business. You will need to detail your long and short term business goals if you are going to equip yourself with the knowledge you need to select the appropriate system.

The key thing to take away is that all phone systems are different. They have different levels of design, varying degrees of integration and specific technological limitations that you cannot ignore.

You do not want to get stuck with a phone system that requires you to punch in a series of complicated codes in order to perform the most basic of functions. Your solution needs to be flexible, powerful and simple.

Your business telephone system needs to support your overall company goals. If your telephone system hinders your business, instead of benefitting it—you have adopted the wrong system.

Look for ease-of-use, accurate features and a dealer that demonstrates your new system in a real world context.

HOW TO BUY A BUSINESS TELEPHONE SYSTEM

A business telephone system can help you run your office more efficiently, reduce costs, improve customer service and enhance your reputation for professionalism. It's likely that 100% of your customers will contact you through your telephone system.

This is one of the most important decisions that you will make for your business. Sales people will try to sell you all the bells and whistles, but you need to look at each feature and decide if it will add value to your business or save you money.

Here are some things you should consider before purchasing your telephone system:

- *What are your telephone system requirements?* Who uses your phones most? How many employees do you have, and how do they communicate with each other and your clients?

- *What system functions do you need?* When someone calls, do you want them to hear an automated message giving options to reach staff members directly or do you prefer to have the receptionist answer the calls? How do you want to receive messages via voicemail, email or text? Do you want call center capabilities? Do you want to integrate your telephone system with your computer network for click to dial and screen popup features?

- *Should you consider VoIP?* Do you have remote or home workers? Do you have multiple locations that you would like to utilize a centralized voicemail system for all of your employees?

- *What about training?* Who will teach my employees how to use the telephone system? Is there system documentation?

- *Request quotes from a few companies, not just one or two.* These quotes must be detailed so that you can make an educated decision about which company is best. Always check the contract fine print! Be sure that the price includes equipment, installation, administration fees, training and warranty information.

- *When you get all of the detailed quotes, sit down and compare the results.* Look at price, warranty and the company's installation processes. Check to see how long the company has been in business, how much experience they have and check their references.

- *Finally, after you have decided on a company to install your new business telephone system*; carefully select installation dates that will not cause a disruption to your business.

THE IMPORTANCE OF CHOOSING THE RIGHT TELEPHONE SYSTEM

Choosing the right telephone system means completely evaluating what your business goals are. What is the future direction of your company? VoIP, voice mail, unified communications, call centers and data are some of the components that you need to consider when choosing a specific system.

The reality is that many companies simply do not have access to a telecom department that would support their chosen system. Businesses should be independent and self-sufficient, or completely able to make changes, add things or move without third party assistance. A good telephone system should be changeable online, so that maintenance and updates can be performed automatically—without any costly interruptions.

More convenient than external equipment in this case, is an interface card that has the in-skin application. These 'in-skin' cards are incredible, allowing companies to add functionality as the need arises. They are also guaranteed that the application will work in conjunction with their chosen telephone system. There are many examples of in-skin cards—the most common being VoIP, conference bridge and voice mail. .

As a basic principle in telephony systems, remember that if it is difficult to use, it will not be used. You can have all the features in the world, but if they are not easy to use, no one will use them. The system 'brain' hides away, but people will be using that telephone on a daily basis.

For successful use of the applications in the system, telephones need to be visually appealing and ergonomically designed for use. There should be a large LCD screen that can be adjusted when used, for displaying important caller IDs and other system information. Users should not have to refer to a user guide all the time, which is why softkeys exist.

The most important function of a telephone system is flexibility, because different businesses have different needs. A business should not have to adapt to their telephone system, or operate in a specific manner due to limited functionality. That's why matching unique business applications to a specific telephone system is key.

All businesses need to understand that choosing the right business telephone system is essential for success. It is how you communicate with customers, affiliates and partners. It can be really frustrating dealing with the process of choosing the right telephone system, but it does not have to be that way!

Luckily, the benefits of choosing the right telephone system far outweigh any research and effort that you put into selecting one.

- The right telephone system can increase productivity and save your company money. If not installed properly, they can cause endless problems.

- Optional features included in your business telephone system give you the edge. An auto-attendant for example saves you the cost of a receptionist and frees you and your staff from having to take every single call, which interrupts your work day.

- Being connected with the right conferencing capabilities can rapidly improve out-of-town business, which will help your business grow at a much faster rate than before. Plus, you will not have to pay for third-party conferencing services.

- If your business relies on sales, you always need someone to answer the phone. With the right system, calls will be bounced to the nearest available staff member, which means you will never lose out on a lead.

- When you take the time to select the best telephone system, it enhances customer service in your company. When your employees are connected, they will be more informed—and with better communication comes smoother operations.

- Telephonic communications often hold a lot of power, in how you are perceived by your customers. When a customer calls in and they get great service, they will be more inclined to buy, or say nice things about your brand online. Reputation and your telephone system go hand-in-hand for this reason.

- The right telephone system also adds value to your company, in the eyes of your employees and customers. When communication works well, and fuels collaboration and productivity—it improves the overall 'value' of your company.

- Choosing the right telephone system allows your business communications to expand unencumbered because it is easily scalable and adaptable. Spending a little more on the best system now, increases your chance of business success as you expand.

10 PITFALLS OF CHOOSING A BUSINESS TELEPHONE SYSTEM

As you can imagine, there are many mistakes that you can make when selecting a business telephone system for your company.

When you choose the wrong system, it can do the complete opposite of what you originally intended it to do. You can end up paying far too much for what are essentially unnecessary 'bells and whistles' and ongoing maintenance.

Not to mention getting snagged into a complicated system that no one can figure out in-house and you cannot expand this system without huge upgrade costs. Be very wary of these 10 common pitfalls when choosing a phone system.

#1: Deciding on a telephone system based on price alone

The most common mistakes business owners make, is that they think all business telephone systems are alike, so they choose one based on price alone. The communications of your company are an operational necessity, so not only does it need to work—but it needs to be comprised of the best possible benefits and features. Cheaper equipment and low cost installations may seem like a good idea, but it rarely is.

#2: Not leaving room for expansion

Any telephone system that you invest in, needs to be able to grow along with your business. You certainly cannot afford the ridiculous upgrade costs every year, as your business grows. Be careful to look to the future (five years or more), and try to plan accordingly. Adding features is possible later on, if you select the right system!

#3: Not training your employees to use your system

An important part of installing a new telephone system means that you will have to train your staff to use it properly. Otherwise, how will it ever work? Invest in the right system, AND invest in training from your phone system vendor.

#4: Paying for continuous tech support

Bottom line, you do not need it. Make sure your telephone system vendor stipulates in your contract that you get free support via phone and email when you need it.

#5: Not choosing the right telephone system vendor

Before any benefits can be reaped, you need to find a vendor that will be able to give you great equipment, great installation, better pricing and limitless support. They need to be there for your company when things go wrong!

#6: Not including your employees in the process

Granted, you are the boss, the business owner. But there are people more qualified to know what you need in terms of telephone systems for your business. Your staff of course! Ask them to contribute a list of what your business needs to have, so that you are sure to get all necessary functionality early. It will save you money instantly.

#7: Neglecting next generation features and capabilities

Business owners with vision will understand that next generation technology is the key to efficient business practices now and in the future. For example, save money by channeling your communications through your computer network utilizing Unified Communications. If you do not have these features now, make a plan to integrate them soon.

#8: If you are a small business go for the easy option

Small business owners often do not want to learn new things. So, they stick to what is cheap and simple to learn. Make sure that you do not just go for what is easy—go for what you NEED.

#9: Ignoring business continuity and recovery

Should your system fail, a plan for recovery should be implemented. Your equipment vendor will help design a plan of action in the event of an outage. Telephone systems can have built in redundancies and failovers.

#10: Not understanding the technology

You would be surprised how trusting business owners can be when buying expensive telephone systems for their company. If you are not informed, then how will you be able to decide which features you need, and which you absolutely do not need?

Take time to understand your business telephone system options. A good first step was buying this book. But when it comes down to choosing your vendor, review these issues and really try to understand the 'why' behind your technology.

CHAPTER 2

Different Types of Telephone Systems

"I'm a great believer that any tool that enhances communication has profound effects in terms of how people can learn from each other, and how they can achieve the kind of freedoms that they're interested in."

- BILL GATES

Designed on different technology platforms, there are several types of business telephone systems to select from. Now that you know how to choose one, all that is left is to gain a better understanding of each option. Which platform will best suit your needs?

KSU-LESS TELEPHONE SYSTEMS.

There are many telephone systems that have the same circuitry that was once housed in a KSU, which was built into the telephones. These systems are called KSU-less systems. KSU-less systems come with both cordless and corded phones. The KSU-less system is simple to install. They are also less expensive, but this means that they might not offer the same features found in systems that utilize a KSU.

Even though these phone systems are affordable, they cannot scale with your company. Typically, they on allow for 2-4 lines and offer basic, novice features. These phones do not need a central control unit, because it is built into the phone itself, which is great for office relocations.

KSU-Less systems are do it yourself retail products sold at Staples and Costco. They do not offer many features like music on hold, line privacy (someone barging into your call accidentally), transferring answered calls into voice mail, and more. They are also built to last 1-3 years. They are not commercial grade quality.

If you have a small company with less than five people that will be using your phones—and you do not plan on expanding, then a KSU-less system is a good choice. It is less expensive than other available options. It is also decentralized, which means that it is totally programmable.

A KSU-less system has comparable features to a smaller commercial telephone system. The key difference is that the communication technology is directly located inside the telephones. There is no need for direct wiring to an office unit and phones are easy to sell, replace or move during office relocation. They are not a permanent on-site investment.

There are however, some limitations with KSU-systems such as the capacity. This system will only work well when the business only needs a few phones and lines. The maximum most KSU-systems can expand to is about 8 phones and 4 lines.

A KSU-system is therefore a great investment, if you do not expect to expand past this limitation. If you do plan on growing your business past this capacity, then you will lose your initial investment as you are forced to upgrade later on.

KEY TELEPHONE SYSTEM

Key Telephone Systems have been around for many years. They are used most often by small to mid-sized businesses with 50 employees or less, located in one building. Offering a variety of features that are commonly needed by small to mid-sized businesses such as intercom, privacy, long distance restrictions, automated attendant and voicemail, conferencing, caller ID and music-on-hold—just to name a few.

A Key telephone system is supported by a key service unit, also called a KSU, which delivers the basic features that you will need to conduct business. An equipment vendor installs and maintains the KSU, which is essentially the brain of your system that gives you the features you need.

These phones come with multiple buttons or keys. Lights are on display, which show the user which lines are in use at the time. To call out, simply press the button to choose the telephone company's central office lines. You will also find buttons like hold, transfer and intercom on these phones.

Once a Key phone system is installed, they are incredibly easy to use and do not require much training. There is a one year warranty on most systems from the manufacturer, and one from the equipment vendor, for service and labor.

The one limitation of Key phone systems is that they lack the advanced features that come with other standard systems, like Unified Communications and Computer Telephone Integration—which perform operations like click to dial and screen pops.

If you only need basic features from your telephone system, and do not need to expand your employee numbers or call volume—then a Key telephone system may be what you need.

PBX SYSTEM

A PBX system, also known as a "Private Branch Exchange" is commonly used by medium to large businesses with 50 employees or more. However, for a small to medium-sized businesses, PBX systems are increasingly valuable as technology develops.

PBX systems connect the internal telephones of independent organizations to the PSTN, or Public Switched Telephone Network, via trunk lines. A company that uses a PBX will need fewer outside lines than extensions, because these extensions won't all be used at once. Extensions are any 'end point' used at the branch, like telephones, modems and fax machines.

They are different from your average Key system in that they automatically choose their outgoing lines, unlike Key systems that do this manually. Hybrid systems (Key and PBX) can contain features of both. A PBX system allows businesses access to flexible system design for their own individual needs—and they can be changed at any time. They are known for both simple and sophisticated features, according to your needs.

Proprietary phones are used by both Key and PBX systems, so users cannot plug other types of phones into these phone jacks. Analog equipment cannot be plugged into digital jacks. A fax machine for example, cannot be directly connected to a jack PBX. The good news is that many PBX manufacturers sell digital to analog adapters.

Plug these adapters into the digital line and you will be able to use analog equipment there, such as credit card machines and fax machines. PBX systems can be easy-to-use and advanced, which supports their complex implementation. There are basic features, like transfers, conference and hold—and advanced features like tenant service, call recording and interactive voice response, and queue groups.

While this system is usually used in businesses with 50 or more employees, it can be used on a smaller scale. They have the greatest flexibility among the business telephone system types, and can support complex additions and implementations as your business grows.

There are many advanced features on a PBX system, and all of them are customizable. Many small businesses that plan on rapid growth invest in these systems, as they can become economical over time—especially if your business grows at a steady rate.

HYBRID TELEPHONE SYSTEMS

This is the most common kind of telephone system sold today. It combines the ease of use and cost of a key system with the expansion and power of a PBX. Hybrid Telephone systems have largely replaced traditional Key and PBX system and come in sizes ranging from 4 phone to over 400. Virutally all hybrid telephone system also come with VoIP capabilities built in, so you can set up remote phones for telecommuters, link multiple offices, or get dial tones directly over the Internet, called SIP trunking. Hybrid systems can even be set up as all VoIP or Pure VoIP systems or in combination with digital and analog phones and phone services.

VOIP: VOICE OVER INTERNET PROTOCOL SYSTEM

PBX systems were significantly improved, thanks to two developments during the 1990's. It gave rise to new types of PBX system. The first development involved the massive growth of data networks and packet switching—which the public was beginning to understand. Businesses needed packet switched networks for data, and using them for phone calls was tempting.

The Internet was now widely available as a global delivery system, which made packet switched communications even more tempting. This eventually led to the development of VoIP technology.

IP Telephony is so much more than simply converting voice signals into data bits. It also involves call set-up messages and networking information, which is visible to the end user. This system makes ease-of-use and convenience possible with these networked telephony systems.

Students at a remote campus, or office workers in a particular department—are able to benefit from the functions and features, like music-on-hold, voice mail, message lights and telephone set displays – though they were connected to a central site.

It is true that IP Telephony is expanding globally, as decision-makers start seriously considering the benefits and potential cost savings that these converged infrastructures offer. VoIP solutions protect your investment, and deliver all the traditional benefits of a PBX system, without the geographical limitations.

Employees working remotely can be added to the company's call centers—so that they can take calls at peak hours during an event, a lunch break or a promotion. This will generate higher call volumes for the on-site call consultants.

The best part is that these remote offices can be located in other time zones, other cities—so that the call centers function 24/7. A single call center can handle calls for the whole company after hours—for service, repair or any other emergency problems. Remote offices mean opportunities for lower cost of living communities and employees that have lower salary requirements. It contains office expenses and reduces employee costs.

VoIP technology includes sophisticated communication features, but they are also easy to scale and robust enough to offer something for every enterprise's needs.

Things to Consider:

A fairly large concern with VoIP systems, involve emergency 911 calls. It is very important to address this issue prior to installing a VoIP System.

VoIP relies on an Internet connection, and the quality of service that you receive will be directly related to the broadband Internet service that you have.

If your PC is faulty, gets a virus or otherwise cannot operate properly—this will also affect your telephone systems. Bad Internet connections or line congestion often results in low quality voice transfer, or distortion. If you simultaneously use your computer while making a VoIP call, sometimes the quality of voice transfer decreases dramatically.

All Internet connections can fail, which in turn affects your business. Consider factors like jitter, packet loss and latency. Transmission errors can affect phone conversations to the point where you cannot even understand the person on the other side anymore. Stability of Internet data transfer must be guaranteed before VoIP completely replaces traditional phone systems.

CLOUD-BASED SYSTEM

Cloud telephone systems are the "newest" product on the business telephone market. Cloud phone systems are hosted PBX systems with a cooler name.

"Cloud" has become the newest buzzword in VoIP and hosted PBX. They change the way you buy communications and refine how you are able to communicate. This allows your workforce to be more flexible, productive and agile—exactly what is needed for business growth.

- Cloud systems provide easy, reliable access to communications for employees around the globe, that need to connect from many locations or areas.

- They allow you to communicate and collaborate from anywhere in the world, at any time of the day or night, from any sort of device – mobile or desktop.

- Cloud systems support remote and mobile employees, which helps them to stay productive when it matters most.

You just need to choose the telephone system that meets all of your communication needs. Keep in mind that cloud systems require a great Internet connection to work. The major downside is that they have the same shortcomings as the Internet does. When the Internet goes down, your telephone system will also be down.

This is why it is important that you have a stable Internet connection. Monitor your Internet speed with a free speed test. This will ensure that your network connection is fast, and can handle a cloud-based system. Typically, you will not have to pay for maintenance, support, or service. Setup and training are incredibly easy.

Your hosted phone system will give you an auto attendant, call forwarding, call waiting, music-on-hold and access to a 'dial-by-name' directory. Using this directory, you can reach anyone on your phone system, instantly.

Look out for:

- Round the clock, 24/7 support and monitoring. This will ensure maximum up-time and will allow your IT staff to work on more important things.

- Redundant servers, so that you don't have any downtime.

The actual Cloud-system is based in the cloud (Internet) and as a subscriber you will connect for a monthly fee.

There are many benefits to using these cloud-based phone systems. The most interesting thing is that you can store your entire phone data in a software program in the cloud, which means it can never be lost or deleted. If you are a small business that simply does not have a large budget for a PBX system or a key system then a cloud-based system may be what you need.

CHAPTER 3

Voice Over Internet Protocol (VoIP)

"Take advantage of every opportunity to practice your communication skills so that when important occasions arise, you will have the gift, the style, the sharpness, the clarity, and the emotions to affect other people."

– JIM ROHN

VOIP is undoubtedly one of the most modern, cost-effective business telephone system options that you are likely to come across in this niche.

Many experts are asking why people still bother using the more traditional forms of paid telephone use. With a VOIP system, your business will never have to worry about paying a high phone bill again.

WHAT IS VOIP?

As we mentioned earlier, VoIP is an acronym for 'voice over Internet protocol.' This is just a fancy way of saying 'phone service over the Internet.'

So, by definition—VoIP is the technology that allows people to make telephone calls over computer networks like the Internet. You use a broadband Internet connection, instead of those traditional analog telephone lines.

At a basic level, you can use interconnected VoIP services to conduct and receive calls, for a nominal service fee.

You also get VoIP systems that involve using a computer, or dedicated a VoIP phone. There are still others that allow you to use landline phones to place VoIP calls, by using special adapters. So there is no one kind of VoIP system, there are several.

If you have decided to use a VoIP system because of the decreased installation and costs, you will want to focus on this chapter to get the most out of your decision.

VoIP is becoming more and more popular for small businesses, because of trends in lower fees and the necessary adoption of fast broadband Internet lines these days. It seems logical to use these faster lines as your dominant method of making calls.

Expect this Internet-based technology to become widespread over the next few years. There are however, some concerns if you are going to opt to use this voice over Internet protocol.

Threats have emerged, as they tend to do with all Internet-based technologies that gain prominence among people. There will be vulnerabilities like scams, attacks and hacking that IP telephony will be prone to once it becomes widely-used.

VOIP CONFIGURATIONS

In any VoIP system there are configurations that you will need which will make up the physical hardware of your business telephone system. They are defined here.

USB Adapters

Use a traditional phone to make VoIP calls by integrating it with a USB adapter. These telephone adapters are usually a little larger than your average USB adapter, but they look close to identical to your average thumb drive.

The adapters come with a modular phone jack that you can manually attach to any average phone line anywhere in the US. The moment you have connected your USB adapter with your phone, it behaves like it is connected to your phone service.

Only you will know that it is really connected to software and the Internet, which allows you to make cheaper VoIP calls.

Dedicated Routers

A dedicated router allows you to use your telephone to make VoIP calls. They are connected to your modem or broadband Internet source, and attach to your standard telephone.

They have to be configured with a VOIP provider and service plan, but once this is done no extra software or computer interaction is needed. It is like having the Internet running through your landline. Just pick up your phone and dial.

Dedicated VoIP Phones

A dedicated VoIP phone looks remarkably like your average telephone, but it is not connected to a phone line—it is connected to a computer network. You will need this phone to have a base station where the actual

Internet connection will be. It can also operate if there is a local wireless connection in the area.

VoIP phone systems also require a service plan with a special VoIP provider. You will pay your basic service fees, but it is much, much cheaper than your standard phone bill.

VoIP Applications or Softphones

In the world today there are tons of software apps (softphones) that let you VoIP call people from your computer. All you need is your headset, microphone, Internet line and soundcard.

These softphones are free to use, and do not cost a penny. The companies that give these softphones away make their money by offering additional Internet telephony services, if you need them. Then you will need to pay a small service fee.

If your clients have this software (like Skype) you can chat to them at absolutely no cost. You can also call landlines and mobiles for a small fee. This setup is ideal for a business that wants to minimize costs across the board.

AVAILABILITY, REQUIREMENTS AND LIMITATIONS

If you are going to consider getting VoIP systems and services— then you need to be aware of the limitations, requirements and availability. There are lots of things to consider before deciding to use this kind of service as your primary means of communication.

Availability and Limitation Concerns

One definite downside of using a service like VoIP, is that during your basic power outage—you are left high and dry, without any phone service at all. VoIP devices like adapters, routers, modems and computers rely on electricity to function.

Without it, your system comes to a grinding halt. Your traditional phone lines however are usually available during a power outage, so business does not stop. There is only one real way to circumvent this limitation—and that is to invest in a UPS.

UPS, or uninterrupted power supply will keep your systems on for a while after the electricity has gone out. This will be a big help in an emergency. Along with your VoIP installation, remember to include a UPS in your desired equipment setup.

You will always require high speed Internet for your VoIP systems to function. The system is only as reliable as your current Internet connection. Even if you have the best high speed Internet money can buy, there will still be service interruptions, degradation of line quality because of high Internet traffic and times when your line is down.

If you attempt to call someone via your VoIP system while there are thousands of people using that same Internet connection—the sound quality could suffer, or you may not be able to connect at all. And then you have calls that drop or disappear because of glitches.

Your basic 911 services are not guaranteed when you use VoIP systems. You can get them with some of the interconnected services that extend VoIP to traditional phone lines however. But, do not assume that 911 works when you run on a VoIP system. Leaving the line open may not lead to anyone coming to help you.

Requirement Concerns

The basic requirements for your VoIP system are a connection to the Internet via your Internet service provider (ISP), a VoIP service plan to extend your calls to traditional landlines and the VOIP software so that you can actually call people.

A traditional phone line however, does not require any of the above to function. Weigh your options carefully before making your decision.

RISKS OF USING VOIP

As mentioned earlier, there are certain risks involved with using VoIP technology, and these risks will only strengthen with time.

Many of these inherent threats are the same you will find for any app or software program online. You will be able to identify most with the spam and phishing abuse that goes on in the world of email systems.

When you use VoIP instead of traditional phone lines, it becomes another pathway for this kind of cyber-abuse, which can lead to a host of trouble for your business. The confidentiality of VoIP conversations is questionable, depending on your configuration and service provider.

Confidential Calls

Many experts have called into question the confidentiality of a VoIP service. You see, VoIP data travels unencrypted over the Internet, and can be stolen. Worst of all, someone may attempt to collect this data to reconstruct a conversation.

There are software programs that exist that can help people do this—they will take bits and pieces of data and reassemble it like a puzzle. Even though this is still a rare occurrence, this activity will only increase as more and more businesses adopt VOIP systems.

Spoofing Explained

Spoofing is when an 'attacker' pretends to be another VoIP caller. This person can inject a fake call ID into your average VoIP call, so that the receiver thinks the call is coming from a known source. Imagine if your known contact, 'your bank' calls you and tricks you into giving them your details.

Things like caller ID's are quite easily replaced with a bit of light hacking. You could get into trouble and be the victim of a number of spoofs involving your social security number, bank accounts or secondary authentication fraud.

This kind of phishing scam can lead to all sorts of identity theft, and you and your employees can end up in a world of regret!

SPIT: Spam Over Internet Telephony

Marketing strategies go hand in hand with all new Internet technology, and VoIP systems are not exempt from that. Telemarketing and spam has been irritating people for decades already. A new hybrid concept called SPIT could see you at the receiving end of some VoIP system spam.

It sends commercial messages via your VoIP system – which is fast and cheap for them, but devastating to your business name. Unsolicited calls can happen, and they do—from your line and to your line. VoIP voicemails alone can require huger amounts of hard drive space.

PROTECTION AGAINST RISKS

After reading all about the risks of VoIP service, it is time you also recognized how to protect against these potential, and growing, risks.

Safe VoIP usage is not difficult, it requires the same vigilance that you use when you are surfing the Internet or managing any number of your other online webpages, apps or programs.

If you ignore these practices, then you open yourself and your clients up to potential hackers and threats every single day. Make sure you implement these into your business policy.

1. Be careful about opening strange files attached to the instant messages that come along with your VoIP system. They could infect your entire system!

2. Create strong, diverse passwords and change them once a month. Do not keep them on your computer, write them in a book and keep them in hard copy in your desk.

3. Always use anti-virus software and anti-spyware programs—keep them updated and run regular scans to clear away unwanted cookies and threats.

4. If you have opted for VoIP software, speak to your provider about using encryption software along with it.

5. Never, ever tell anyone your personal information. That means if a friend asks you, assume that they are a hacker and will steal your identity or money. Under no circumstances will anyone decent ever need to know these things about you.

6. Configure your web browser security so that hackers cannot gain access to your PC, where they could seize control of your VOIP systems. This can lead to data theft.

7. Always use a good firewall, if you do not have one—buy one. If your provider tells you that you need something better (a paid firewall) then invest in one.

8. Verify the security and authenticity of new software and files that you have just downloaded. You can scan them with your anti-virus software.

9. If you have personal data on your computer, like financial records then back it up and password protect it on your PC.

10. Always update your apps and software. When you leave them and do not update, hackers can use these gaps to let themselves into your computer.

VOIP STANDARDS AND PROTOCOLS

VoIP systems use Internet Protocol to send voice 'packets' over an IP network. This means the VoIP can be used on ANY data network that uses IP.

Not only does VoIP work on the Internet, it can also work on intranets and local area networks. How does it work? Your voice signal is first digitized, compressed and converted into an IP packet.

This packet is then transmitted over the IP network. Signaling protocols are used to create and close calls, send and receive information, locate users and negotiate capabilities. This is straightforward, simple and it is a very low cost system model.

There are other reasons why a small or medium sized business might want to use VoIP systems, instead of the more traditional phone systems.

The increasing demand for integrated multimedia communication and the growing interest in voice and data network integration are two other reasons why a business owner may prefer VoIP systems.

Even though IP was originally designed to transport data, a few issues have cropped up in VoIP systems because of vendors not conforming to traditional standards of quality.

Interoperability for example can be seen as an issue. On public networks, products from different vendors need to operate with each other if VoIP is going to become popular among business users. Standards need to be devised and adhered to across the board.

The quality of a person's voice is also in question—as IP carries data, but no guarantees that voice quality will remain intact over the network. There may be delays, distortion and echoes. Packet prioritization has been looked at, as a standard solution.

Integration with PST networks has also been proposed. While VoIP technology is being adopted, it needs to be able to work in conjunction with the older PSTN technology. In other words PSTN and IP networks need to look like the same thing to new users.

Security standards are a concern, as people may be able to capture packets meant for someone else. Security encryption and tunneling must be devised to protect the information that moves across the IP network.

Finally, scalability will become a main issue with regards to sorting out VoIP standards and protocols. Quality needs to be the same for IP calls, even though they are much cheaper. VoIP systems will gain popularity if these standards and protocols are implemented.

CHAPTER 4

Cloud Based Telephone Systems (Hosted)

"I'm a great believer that any tool that enhances communication has profound effects in terms of how people can learn from each other and how they can achieve the kind of freedoms that they're interested in."

- BILL GATES

A cloud based telephone system is hosted, much like VoIP PBX systems are. This particular system is also called an 'Internet phone system' as it is permanently connected to the Internet. In business, few modern systems are as cost-effective or multifunctional as a cloud based telephone system.

The data that is used in the system is stored on a secure server that is accessible online. This means that you can access your business telephone information anywhere, and it will never get lost due to power outages or equipment failures.

WHAT IS A CLOUD BASED SYSTEM?

A hosted cloud based telephone system is ideal for small to medium sized businesses. It gives you all the benefits of a large scale business communications system on a small business budget.

These systems are replacing the PBX phone systems for smaller companies, and they are far simpler in many ways. No thick manual is required to learn how to use the system, and there aren't dozens of wires that need to be run all around your office.

Cloud based systems allow big businesses to save money, and they provide an affordable option for smaller businesses that do not have the time or money to spend on complex telephone solutions.

These systems put users in control, as they have just as many adaptable features as hosted PBX systems—only without the wiring, additional costs, and need for employee training. Cloud based systems give smaller businesses access to automatic receptionists, voicemail boxes, call transfers, smart call routing, and easy-to-use extension structures.

On the whole, these systems do a lot to improve the business image of the company, and with the collective features, productivity nearly always improves. As a business, you can choose which features you need on a selective basis.

These hosted cloud based systems are leading the way in modern telecommunications. Even though phone costs are on the rise, it's great to know that there are alternative solutions for businesses with a limited budget. For a scalable, flexible telephone system that is also simple to operate, your business should consider getting a hosted cloud based system.

ADVANTAGES OF USING A CLOUD BASED SYSTEM

There are many advantages to using hosted cloud based systems. It is these benefits that make this system the most logical solution for small to medium sized businesses today. For a better understanding of why this kind of system would work best for your business, here are the main advantages of installing a hosted cloud based system.

- **Automatic upgrades:** Your hosted cloud based system will automatically upgrade itself, which means that you do not need to pay an expensive consultant to do it for you. There are no maintenance fees, and your system will always be up to date.

- **Fully redundant network:** A good hosted cloud based system will have a fully redundant network, which means you will always have a dial tone. You will not have to worry about down time, disconnection, or lost business.

- **Cost-efficient:** Hosted cloud based systems are very cost-effective as they require no capital for use. You do not have to buy equipment such as telephones or pay for expensive installs or upgrades. With no maintenance costs or start-up capital, any small business can afford to add a cloud based system to their business. But keep in mind that you almost always have to buy the phones or you just end up paying extra to rent the phones. Plus POE switches and QoS routers are required. Also, you pay per phone rather than per outside line, so if you need lots of phones, but don't make that many concurrent phone calls the overall cost is much higher. This is the main reason most companies aren't right for hosted systems. The average company has 1 line for every 2 or 3 phones, so the added monthly cost of paying per phone vs per line makes hosted a bad investment for some. Additionally, make cloud based companies sell phones that are locked and can only be used on their service. Unlocked phones are more expensive.

- **Hosted support:** If something goes wrong with your system that you cannot fix on your own, your hosting company will fix the problem off-site. In this way, your system is repaired remotely and does not interrupt the flow of business. If the problem is in your network or your wiring you need a local vendor to fix the problem.

- **Increased flexibility:** Every business is different, and your telephone system should fit in with the way you run your business on a daily basis. Use the web portal to make changes to your system as you need them. There is no need for IT staff; simply make the changes yourself. The tools are yours to tweak and refine.

- **Increased scalability:** With hosted cloud based systems you do not need wiring skills and a budget to add and remove features. If you need an additional phone, just plug it in and sync it to your system. If you need more functionality, just add it from your web portal dashboard. In this way your phone system grows with you.

LIMITATIONS OF USING A CLOUD BASED SYSTEM

A hosted cloud based system may be the most affordable and scalable of the current telephone systems on the market, but it does have its limitations. It is important to understand these before you invest in this type of system, as it can affect the way your business works.

- **High speed internet access is critical:** Your ability to connect with customers and suppliers depends on how fast and reliable your internet lines are. You can only use hosted cloud based systems on fast lines, and while these are becoming more prolific—the Internet can turn off for periods of time.

- **Access to emergency services:** You access to emergency services is limited, as you will not be using normal phone lines to call for certain services. Make sure you let the operator know where you are, as they are unable to trace your call when it is routed through a hosted cloud based system.

- **Natural interferences:** Storms, fires and other natural disasters can interfere with your internet line speed or quality. This means that your calls could sometimes drop, or be disconnected, and you may experience trouble during the periods of time your internet is completely down.

- **Decreased sound quality:** Depending on the line, you may experience sound quality interference or distortions. This happens when outside factors make it difficult for packets of information to pass from one system to another.

- **Reliant on electricity:** When you use hosted cloud based systems you become reliant on electricity. To keep your phones working, you must install a UPS, or uninterrupted power supply, to charge your systems when the electricity is out.

- **Waiting period for problems to be fixed:** When something big happens to your cloud based phone system, it can take a few hours to fix. Your host will let you know that they are working on the problem, and they will notify you when they are done. The support is great, but it can lead to loss of sales.

These limitations do not mean that hosted cloud based systems are no good, but like any system—it has its weak points. Most small and medium sized businesses can easily work around these limitations so that they can enjoy the benefits of owning a scalable, flexible, and cost-effective system like this.

IMPLEMENTATION OF A CLOUD BASED SYSTEM

The first step towards implementing a hosted cloud based system is to find a company that you trust that has the system you want. These companies offer different packages according to your needs, and they will charge you a monthly fee for using them.

Once you have signed up with your chosen provider, they will provide you with the hardware or software that you need to get setup. This could be a basic IP phone, or a program that you can download for use. These 'end points,' as they are called, use the network facilities of the host provider as a gateway.

The bulk of your services are off-site, which means that whenever something goes wrong with your cloud based system—you can call it in, email or otherwise contact your host for off-site help. They will remotely access your system and repair the issue from their current location. In this way, there are no maintenance fees or repair men to slow business.

The upfront costs involved with hosted cloud based systems are minimal. You do not have to buy the phone system outright; it is leased from your host for a small monthly fee. Over time, you will save money because calls will be cheaper and no capital outlay or paid on-going support was needed.

Once you have signed up with your host, they will pre-configure your phone settings for you, based on your needs. After installing or downloading the software, all you need to do it plug in your IP phone and you are ready to go.

At this point, you can activate or remove features that you will use by logging into your web portal. Once there, you will be able to browse lists of features that may come in handy for your business. Simply eliminate the ones you do not need, and activate the ones you do. Standard moves, changes and add-ons take no time at all.

Once you have setup your hosted cloud based system the way you like it, any of your employees will be able to use it. There is almost no training required, as the service walks you through the basic processes and features as you use them.

Above all, this phone system is user-friendly. You will not find an easier setup for any other kind of business telephone system, which makes moving locations a simple process. All in all, it would take a couple hours of your day to get this system up and running.

CHAPTER 5

Voicemail / Automated Attendant

"Take advantage of every opportunity to practice your communication skills so that when important occasions arise, you will have the gift, the style, the sharpness, the clarity, and the emotions to affect other people."

- JIM ROHN

Voicemail has been around since the '70s, but it was Gordon Mathews who patented a voicemail service in 1982. His company, VMX, provided the first voicemail services in the world. They recorded and managed messages using the digital technology of that time.

Since then, voicemail and automated attendant have evolved. Currently, voicemails are digital recordings of outgoing and incoming messages. These messages are managed by an on-site or an off-site system.

WHAT IS VOICEMAIL?

Voicemail, also called 'phone mail' or 'VMS' is usually a centralized system for storing telephone messages. The recipient can later retrieve these messages for review and response. Voicemail is also a term used to describe the system of conveying a stored voice message using hardware such as an answering machine.

Different voicemail services offer different features. Businesses can buy systems that can then be managed by their own staff, or they can have a contract with a third party company. It works particularly well with hosted VoIP or cloud based systems, as a simple program can transform a PC into an electronic messaging system.

Voicemail allows your company to be more flexible as information flows between customers, suppliers and staff—even when both parties are not present for the exchange. Businesses rely on voicemail to manage customer expectations and to improve customer service. Leaving and sending messages is key to any quality customer service strategy.

These voicemail systems can be integrated with an in-house phone system, which makes a whole host of new features available to the end users. A few features that voicemail offers are urgent message delivery, paging, and off-site access to messages.

In the early days of voicemail, real people used to be the recipients of your message. These days, however, these voicemail systems come with a built-in operator. Known as auto-attendants, these programs guide users and help them make the best out of their experience with that voice mailbox.

There are several different types of voicemail available to your business. Certain systems are designed for specific use, but it depends on who is using it, how, what equipment is being utilized, and who is providing the service.

Voice mail to Email Integration is one of the most popular features on modern voice mail system. You get your voice mail sent to your inbox or smartphone as an audio file attachment. No more calling in to check messages and messages are delivered almost instantly.

WHAT IS AUTOMATED ATTENDANT

An automated attendant, or a digital receptionist, comes as part of a business telephone system that contains voicemail. It will automatically greet and direct your caller using an automated message, and it will then route the call as needed or record a message.

There are menu options that the caller needs to work through in order to achieve the desired result. By pressing on the right keys, or speaking particular words, the automated attendant can organize and store (or direct) your message in a way that is easiest for you.

These days automated attendants have become very smart, and are able to route multiple simultaneous phone calls to the right people at once. They are commonly used as receptionists, if your receptionist is on lunch, out, or not available.

By using an automated attendant you guarantee that no crucial messages are missed and that your business can function at full capacity—even when you are not there. It certainly speeds response times, which naturally improves customer service overall.

End users can manually turn the auto attendant on and off as needed. Some companies prefer to only use the call forwarding features; others make full use of auto attendants when they are very busy.

These voice menu systems mean the caller can always get through to your business, which means that you lose fewer customers because of frustration. If a caller wants to find a user on the phone system, they only need to use the dial-a-name directory.

The auto attendant lists users by name, so if you need to get through to a specific person or department there will always be someone available to take your call. If, however, all the end users are not available, the auto attendant will instruct you to leave a message.

The next time the end user wants to check messages, he or she will receive all the customer or supplier calls that were missed during the day. This is a simple matter of reviewing the message by phone, pager, or PC desktop.

Reviewing messages like this can usually be done remotely from anywhere, from any device. Smartphones, tablets, and PDAs can be connected to your system to allow for easy message retrieval. It makes it easier for you to do your job as a business owner.

VOICEMAIL FEATURES

As I mentioned before, the features on your voicemail system depend largely on who you have decided to use as a host and what they have offered you. The good news is that most voicemail services are fairly standard, so you will not miss much if you just get the basic voicemail and attendant functionality.

- **Out of office greeting:** Set one or more out-of-office greetings for your customers and suppliers when you are not around to take their calls.

- **Call back or reply:** As you listen to your voicemails, you can reply to the sender or go directly to their voicemail system without actually talking to them. You can also call the person back automatically without having to hang up and dial the number from scratch.

- **Deleted message retrieval:** During a session, you can restore deleted messages to your system. That means if you delete something by mistake, you can undelete it—as long as you do not hang up first.

- **Message identifier:** Before every message, you will hear the date, time, and phone number of the call—which makes it easier to focus on what the message says.

- **Urgent message retrieval:** Order your messages however you want them; listen to them as urgent, new, or old messages.

- **Outside caller messages:** Choose to listen to outside caller messages first, before listening to internal calls. When an outside caller phones in, you will be able to see who it is, thanks to caller ID.

- **Passwords and security:** Business messages are kept private with special passwords that are used to keep your information private. All passwords need to be highly secure, so easy password options are not available.

- **Management features:** There are many other management features that come with voicemail. You will be able to set extended absence greetings, manage your mailbox, your playback lists, and general commands.

AUTOMATED ATTENDANT FEATURES

Automated attendants may be part of a voicemail system, but they are multidimensional on their own. Imagine all the things a person could do to manage your calls. The auto-attendant does it better. It can be used as a standalone or multilevel configuration, depending on the needs and size of your business.

- **Multiple phone lines:** Your automated attendant will answer multiple phone lines for you, and it will transfer your calls to the correct department or staff member.

- **Direct messaging:** Your auto-attendant will transfer callers directly into your voice mailbox—either with or without the standard greeting.

- **Menus and submenus:** The auto-attendant will present your callers with a list of menu options to choose from that will help them place their call. In large businesses this can be quite a complex process, but the attendant makes it easy for the customer.

- **Different greetings:** Set one or more greetings, at different times of the day. You can also set themed greetings to play during holiday seasons.

- **Frequently asked questions:** Your automated attendant will play pre-recorded FAQs, or frequently asked questions, to customers that call in looking for solutions that are already outlined in these recordings.

- **Directory of staff:** The attendant will play a directory of staff phone numbers that allows callers to be transferred to any extension they enter.

- **Non-reachable extensions:** The auto-attendant can be programmed with a list of extensions that callers will not be able to reach at any time.

- **Cell phone transfer:** The auto-attendant will be able to provide select callers with access to staff members, by connecting the caller to the user's cell phone when they are not on premises.

- **Camp on call back:** If the line is busy, the attendant can put you on 'camp,' until a line becomes free. Then the valid caller ID will kick in, and the customer's number will be dialed to return the call.

- **Fax detection:** Among the many other features is fax detection, as the auto-attendant can recognize a fax tone. It will automatically redirect the fax to the appropriate department or person.

ADVANTAGES OF USING VOICEMAIL/ AUTOMATED ATTENDANT

There are many advantages to using voicemail and automated attendants in your business. They are crucial parts of your overall business telephone system and bridge the gap that is created when you cannot tend to customer needs. Voicemail is the most basic application that every growing company needs in their phone infrastructure.

- **Customer response time improvement:** With voicemail and automated attendants, you will not miss out on business or important calls when you are not around. In this way you improve customer response times and brand loyalty grows.

- **No need for a receptionist:** When you are a small company, an auto-attendant is far cheaper to run than hiring an actual receptionist. If your needs are basic, then this program will ensure that your business moves forward.

- **Fast message delivery:** Speed can mean everything to a growing business. If an emergency occurs and you are not in at the time, you will still receive the messages instantly and be notified of the problem. You can then handle it in due time.

- **Easy message review:** Voicemails are accurate, organized and easy to use. Simply play them back, bookmark them or have them transcribed on your PC for your records. They prevent mistakes and guarantee accurate record keeping.

- **Access to messages anywhere:** Because voicemail can work in the cloud with your business telephone system, you can access your messages anywhere. Whether you are in another country or out for lunch—you will always get the message.

- **Personal touches:** You can choose to use the standard greetings or record some personal greetings of your own. This adds a nice touch to your service profile, and makes your customers feel at home.

- **Voicemail to email:** You can send your important voicemails to your email inbox so that they can be sent elsewhere or recorded.
- **Permanent caller ID:** Your auto-attendant will make sure that you only receive calls from credible sources. No spam calls, marketing calls, or waste-of-time calls will get through to bother you.
- **Time saving:** This system tends to save you a lot of time and effort in the field of telecommunication. When you save time, you save money and can attend to your callers when you get the chance.

IMPLEMENTATION

If you are going to implement voicemail into your business telephone system, you will need to make sure that it comes as an option or added feature when you sign up. Having a telephone system with voicemail greatly reduces the cost of business communication in general. The benefits are worth ensuring that your system has added voicemail.

When you add voicemail to your business, it gives it an instant professional boost and many advanced features that will help you grow. Check with your provider how the installation of your new voicemail system will work.

It needs to be aligned with certain computer specifications, server support, and operational requirements. If you hire a third part voicemail company to handle your mails, then expect to receive a document detailing how the installation and setup process will work.

These companies are very supportive, and they will walk you through the process and initial settings. Everything from checking the amount of disk space on your computer to installing the various program features will be done initially.

Web based voicemail works in conjunction with your existing business telephone system to create a service that is easy to change to your specific needs. All you have to do is find a quality provider and make sure that all systems are compatible.

Once you have installed or downloaded the system, you will set your preferences and automate certain processes such as monthly 'clean-outs' or message saving. The company you choose to work with will then test your voicemail and automated attendant services to check that they are operating correctly.

This system should update automatically, especially if it's linked to the Internet. We suggest using a business telephone system provider to source your required voicemail so that you do not have to deal with multiple providers for a single communications service.

Once your voicemail service is up and running, you will quickly discover what you need and what you do not need. Login to your web portal and remove the features you do not need. Look around for other features that you can activate that will be good for your business image.

MANAGING YOUR MESSAGES

There are a number of ways that you can manage your voicemail messages. These policies will need to be decided once you begin using your new system. Keep in mind that messages will automatically delete themselves after a certain amount of time, unless you specify otherwise. Messages must be managed in a systematic way on a dashboard.

Your voicemail service will come with a web portal where you can log in and manage your messages in a professional online workspace. Sort your messages, listen to them, delete the ones that you have taken care of, and save the messages you want to keep as records.

- Determine how long your voicemail system retains messages.
- Decide which messages you should save as records (in case of disputes).
- Convert certain messages into emails or transcripts for printing and sending.
- Queue up the messages you need to respond to, and respond.
- Add notes to your messages that will help you sort out the query or concern.
- Keep sound files of every message as a record.

It is up to you how you want to manage your messages on your new voicemail system. The key takeaway here is that you definitely want a dashboard on your PC, or a place where you can work with your messages on a more advanced level than simply 'redialing and responding.' Modern systems call for speed, accuracy, and ease-of-use.

When you get started with your voicemail system, you will also learn how to manage messages direct from your phone. There, you will be able to delete, auto-dial, replay messages, fast forward or rewind them, bookmark message areas, and much more.

It is normal for all played messages to automatically delete themselves after seven days. Always make sure that you turn off this feature if it does not correspond with your business model. Message management should be scalable, flexible, and available anywhere you are. That is why it is important to have an online dashboard in conjunction with your phone management settings.

When you get a chance, you will be able to choose the features that you want in your business. Receive wireless notifications, get pager notifications, and prepare group mailboxes for your staff. Remember, you will not be the only one using this message system, so it needs to be really user-friendly.

CHAPTER 6

Your Guide To Unified Communications

"Communication is a skill that you can learn. It's like riding a bicycle or typing. If you're willing to work at it, you can rapidly improve the quality of every part of your life."

- BRIAN TRACY

IP telephony is the foundation for the requirements that need to be met with unified communications in business. Unified communications, or UC as it is commonly called, attempts to provide instant communication between staff that goes beyond traditional devices such as phone and email.

In fact, it serves to unify additional types of communication media including wireless, video, instant messaging, and voice recording. What started out as a trend has now become big business as companies all over the globe are switching to unified communication services.

WHAT IS UNIFIED COMMUNICATIONS?

Unified communications is all about bringing together modern forms of communication –apps, data, video, and audio. The goal of this technology is to provide easy access to data for employees who work in a single environment. It makes sense that to get the most out of your communication efforts, all media should be networked, connected, and integrated.

But you can't BUY unified communications because it's just a theory. Integrating different kinds of communication media to enhance business processes is not a new concept. But it is becoming more important as time moves forward. While you cannot buy unified communications, you can have a unified communications strategy.

A dedicated business telephone systems provider will be able to help you design, develop and build a unified communications system that will complement your business practices. Instead of choosing a simple telephone system, you will have the opportunity to streamline, converge, and network your communication pathways together.

Well planned UC will improve communications and collaboration between customers, staff, and owners. Ideally, these unified communications will keep people in touch all the time through a range of wireless devices or applications that have no central location. You can use or access them anywhere!

Unified communication tries to boost collaboration wherever a need for it exists. When people are more connected, they are easier to reach, and customer service is greatly improved. Things that usually impede business such as travel, technology break down, and human error are surpassed with a unified communication system.

BENEFITS OF UC

There are dozens of benefits to implementing unified communications on your business premises. Above all, the chance to update your telephony to the point where all modern media can be used to improve your business is not something to ignore! Done correctly, UC could be the solution that you have been looking for.

- **UC boosts productivity and innovation:** When all of your communication can be managed and accessed from a single location, it improves productivity and lends itself to future innovation due to the increase in speed of delivery.

- **Elimination of traditional communication silos:** When communication is difficult, disconnected, and hard to use, the business suffers. Having 10 different forms of communication with no unified process or system in place results in confusion, delay, and repetition.

- **Collaboration:** Easily one of the best benefits of UC is collaboration. When sharing data is done collaboratively across multiple devices, people are always informed. Ideas and data are swapped and shared instantly and processes speed up.

- **Increased communication:** When all available devices and systems are under one roof, communication improves. Changing a voice message into an email or a text message has never been easier. With increased communication comes more efficient day-to-day strategies.

- **Ease of access:** Clearly a prominent benefit is ease of access. When all staff can connect to each other, and the business, in multiple ways, things happen faster. Work is done quicker and progress is made on a more measurable level.

- **Instant information:** When you have access to everyone in your business' current availability and presence, you are better able to respond to customer needs.

- **More customers:** When communication becomes easy, so does nurturing quality relationships with customers. Your business will grow because you have more time to build solid foundations with the customers you have, while simultaneously finding new customers.

These business benefits of unified communications are just the tip of the iceberg. In fact, there are many far reaching implications of integrating all of your communication systems together. Your bottom line, for example, will almost always benefit when a communication practice improves.

There is certain continuity in your business processes when you implement unified communications in your company. Things such as supply chain management and customer relationship management will also receive a healthy boost from the improvement in communication strategy.

CAN IT REALLY INCREASE PRODUCTIVITY?

Building business productivity can be one of the main challenges any small to medium sized business can face. The good news is that unified communications is a trend that has really caught on, and it is increasing productivity for businesses of all sizes across the globe.

When you streamline your communication processes, you are doing it for one singular reason—to improve the overall efficiency of information sharing. This means that all elements – people, location, devices, systems, and processes - are affected.

If all forms of communication are connected, they can be easily integrated with each other. Your business will be able to respond to its customers in an unparalleled variety of ways. A customer, for example, may call and leave you a voice message. In response you can choose to email, call, IM, text message, or video conference with them.

It really depends on your needs. Great options means more optimized processes. For example, if that customer was requesting a pricelist, in the past it would be an absolute debacle getting them the right information that would improve sales. With unified communication, a simple click and send is enough.

Productivity naturally increases as time is saved, resources are conserved, and costs are kept to a bare minimum. Business owners understand that if they can regularly achieve these three basic features of business, productivity will increase. This is because there will be more money, more time, and happier customers dealing with your company.

There is real power in convergent technology, especially when it improves customer-facing processes and internal functions of the business. In a recent study by Sonus Networks,[1] they concluded that unified communication has a very important role to play in productivity.

What you can take away from this study is that the larger a company gets, the more time they spend on filtering incoming communications. A basic UC infrastructure can improve productivity lost on ineffective communication by as must as 23%, or up to 1.21 hours a day per employee.

If your small or medium sized business is looking for a way to rapidly improve your chances of annual growth and income, you need to consider getting yourself a unified communications infrastructure.

---

1. Tworek, Richard, Unified Communications Productivity Study $13,000/year, http://qovia.com/2013/03/08/unified-communications-productivity-study/

IMPLEMENTATION CONCERNS

There are some concerns that you need to be aware of when implementing unified communications infrastructure into your current business. We already know that UC gives your business the right kind of communication using the right devices. These are accessible to anyone that works in your company. As a result, certain challenges arise that need to be dealt with before implementation.

- **Cultural challenges:** When UC is first implemented in a company, there are those people who embrace it and those who reject it. It takes a significant cultural shift to get everyone using the technology correctly. This means keeping eager employees from overusing it and helping hesitant employees learn what it is about.

- **Policy challenges:** You will need to make it clear to your employees that in order for UC to work, they need to be open to other forms of customer communication. Most people are not yet excited about the idea that customers can call their cell phones, leave text messages, or video conference whenever they feel like it.

- **Communication interruptions:** Having this speedy form of communication will be rough, because it will escalate over time. This means that your employees need to be prepared to be interrupted many more times a day, and they need to learn how to recover from interruption in a much faster manner.

- **Education challenges:** In order for your staff to properly use unified communications, they need to understand which medium is the best to use at that predefined moment. The only way they can learn this is if you take the time to teach them.

- **Baby boomer challengers:** The baby boomer generation is always reluctant to try out new technologies or systems. They usually do not understand the concept or premise behind unified communications, so spend extra time making sure they do.

- **Abuse challenges:** Some employees will abuse your unified communications system. You can, however, put monitoring and control software in place to curb the abuse of technologies such as text messages, IMs, and video conferencing.

If you can deal with the human side of integrating these new systems, you will not have many concerns to deal with. There is always a teething phase whenever something new is implemented in business, so prepare for it. You will have to learn how these integrated systems work, alongside your employees.

FEATURES OF UNIFIED COMMUNICATIONS

Unified communication is the most advanced infrastructure you can hope for in our modern times. As a result, it contains some of the most in-demand and highly useful features that will benefit any business owner and staff. Consider these incredible features before deciding on your communication infrastructure.

- **Unified messaging:** There is one inbox available per employee where all messages are kept. This includes voicemails, video messages, faxes, and emails.

- **Integrated conferencing:** Unified communications allow employees and customers to host web or audio conferencing chats. This is important for improving customer relationships and ensuring repeat sales, especially with overseas clients.

- **One number reach:** Every individual will have a number that you can use to reach them. Simply enter the number into the phone and the call, message, or media will automatically be redirected to their phone or message box.

- **Presence:** At all times within your company, you will be able to tell who is and who is not available to help a client in need.

- **Disaster recovery:** If business is impeded by a natural disaster, all calls and messages can be routed to another location. In this way your business does not lose any customers or sales due to Mother Nature.

- **Screen pops with customer information:** With UC you can keep profiles on all customers to improve relationships. When a customer calls, a screen will pop up informing you of their past correspondence, connection to your business, name, and personal details. This goes a long way when keeping clients happy.

- **Instant messaging:** There are times during the day when you are in meetings, listening to a speech, or otherwise indisposed. At these times, instant messaging is the quietest and most non-intrusive method of being contacted.

- **Extension dialing:** Your UC will automatically dial the number you want to reach, regardless of domestic or international extensions. No more scratching around looking for the area code in Venezuela.

- **Click to dial features:** You get click to dial features in your email box or from your web browser. It makes calling people easier and allows for cross communication during the day.

Unified communications also manages your contacts more efficiently, and they are available to relevant people within your business. Using this infrastructure you will be able to ensure the sustainable expansion of your company with modern communication on your side.

CHAPTER 7

Call Accounting Defined

"Face-to-face communication is difficult to maintain, and as such, we are continually becoming more reliant on other, easier forms of communication to maintain these relationships."

- BRAD HUNTER

One of the most important facets of business telephone systems is call accounting, yet it often goes unnoticed in smaller business models. Call accounting provides your business with the analytics data it needs to improve business communications on a business-wide scale. All business owners should consider this a vital part of their systems.

Used correctly, call accounting can help you prevent communication abuses by employees in your company, and it can lead to a reduction in costs across many of the core types of communication that your business uses on a regular basis.

WHAT IS CALL ACCOUNTING?

Call accounting is a useful software program that gathers and records all of your telephone usage. It should be part of a unified communications infrastructure as a method of tracking behaviors and changing them to suit the needs of the business.

Call accounting usually provides the user with a historical record of all incoming, outgoing, and internal calls. Depending on the program, it can have many other features that are useful for analytics measurement as well, such as call costing, alarms, and billing applications.

The entire system works by tapping into your telephonic data directly from your communication systems. It then translates this bulk information into something you can use as a business owner. These are called 'insights.' You can use these analytics insights to alter the policies in your business or change the behavior of your staff.

If people have been calling too much, you can reduce the amount of time that each staff member is allowed to remain on the phone. If a certain method of calling clients is working like a charm, that can be implemented company-wide.

Your PBX or telephone system creates things called 'call detail records' for every call that is received. This information is then stored in a server and processed by the software. The goals of call accounting are to calculate the cost of telephone calls and to allocate the cost to particular users.

You will gain new insights into your telephonic communications so that you can measure and track employee performance, and improve communications overall. It is a very valuable way to consistently improve your business telephone system practices.

BENEFITS OF ACCURATE CALL ACCOUNTING

Call accounting has some very powerful benefits for your company. Saving money on telecommunication expenses is key to a successful business model, and this means you will need to track and understand your company's usage. When you have accurate call accounting on your side, you will not have to worry about keeping things in check.

- **Optimize your rates:** With call accounting, you can aggregate call expenses by type and carrier, which means that you can negotiate better rates with certain service providers. You can also institute new policies to help manage usage.

- **Accurate cost allocation:** Perhaps the most important use for call accounting is to accurately allocate telephone expenses to the right department or employee. This helps when you're evaluating human performance as you can see who has spent the most on calls.

- **Reduced management time:** Call accounting replaces manual tracking of telephone usage, which is an arduous process and difficult to get right without software to help you. With the right software, however, you can spend your time making decisions instead of trying to work out who did what when.

- **Network and resource optimization:** Call accounting can help your business identify a wide range of improvements that are needed—such as data delivery capacity and call routing efficiency. Using this data, you can streamline every physical process.

- **Educational tools:** Because call accounting software produces useable reports, you can use them as educational tools to help managers in your company come up with better solutions for staff behavior or discipline.

- **Support available 24/7:** When you get accurate call accounting software from the same provider that setup your unified communications infrastructure, you get round the clock support. Installation, setup, and fixes are available 24/7.

- **Improves employee production:** Eliminate unnecessary calling and make your employees accountable for their telephone usage while at work. Fewer calls mean lighter bills at the end of the month.

- **Contain costs:** Above all, call accounting serves to contain the telephonic costs of your business. It is incredible how much money can be wasted on calls if they are left unmonitored. Make sure that you always know how your communication systems are being used by your staff.

Call accounting is beneficial to everyone. It helps business owners set better policy regarding telephone usage, and it decreases monthly call costs. For employees, it helps regulate how much time they spend on the phone and allows them to come up with solutions on how to reduce this time.

FEATURES OF CALL ACCOUNTING

There are many features of call accounting that would interest you. The first thing you need to know is that not all call accounting programs are alike. Some are very basic, while others are more comprehensive. You will find the SaaS or Software as a Service companies usually provide the better call accounting software for your business.

- **Call record storage**: Call accounting systems can store thousands of internal calls, which you can use to track employee behavior.

- **Flexible call costing**: Everything from capped call costs, customer charging rates, flag fall, and free, local, STD, and international rates are tracked and monitored. It's entirely user configurable.

- **Direct upload**: Many call accounting programs are able to use interface software to directly link with your telephony systems to upload relevant data.

- **User profile**: The user profile feature links call accounting, LAN inventory, directory, and cable management functions together.

- **Tracking and billing**: Call accounting software tracks and bills all calls, including local, long distance, incoming, operator assisted, internal, and international.

- **Reports**: Call accounting software automatically creates reports for you based on custom formats and predefined system information. This is then sent to your email, or you will be notified of the report.

- **Auto-delete function**: There is an auto-delete function that manages old call logs, which prevents the hard drive from filling up.

- **Multi user**: Many call accounting programs are accessible by more than one person so that billing and analytics can be handled efficiently.

- **Password protected**: The system is password protected so that none of your business data can be accessed by any staff members without the code.

- **Real-time call costing**: As calls are made, the system will add the new data to a daily report that it is compiling. You will be able to see who called, when, from where, and how long they were on the phone.

If you decide to invest in some quality call accounting software, then you will not have to struggle to track and monitor employee behavior on an intrusive level. Instead, you will be able to review these reports, think about solutions and address problems —before they recur.

TEN REASONS WHY YOU NEED CALL ACCOUNTING

If you still are not convinced that call accounting is necessary for your growing business, then you will want to consider the following reasons why you need call accounting. These are just some of the reasons why business owners all over the globe decide to implement this software into their communication systems from day one.

#1: It saves you time: Tracking calls manually is very labor intensive and unnecessary. If your employees are still manually recording calls and matching them to phone bills, then you need call accounting more than ever.

#2: It saves you money: When you are able to see where the problems are, you can automatically put strategies in place to fix them. If one department is calling too much, you can put a stop to it or find out why it is happening.

#3: Increases productivity: When employees know what is expected of them, they are able to improve in their job. Not having to manually record calls leaves more time for them to get on with things that really matter.

#4: Budget control: Telecommunication costs are one of the key areas of expense within your business. For small businesses, it matters how you control your telephone budget.

#5: Promotes innovation: When you read the reports and monitor the usage, your staff will be motivated to devise innovative ways to improve the way they interact with telephones.

#6: Improved management: Using these call reports, you will know when some employees are not calling and others are over-calling. It makes you a better manager when you know what is going on.

#7: Marketing results: See if a new promotion works by measuring how many calls you get that week against your usual numbers. Try and increase interest in certain products, and use call accounting to see if it works.

#8: Performance: If you have telephonic departments in your business, such as a call center or help line, you will be able to see if they are doing their job. If they are not meeting minimum quotas, you will know why.

#9: Employee accountability: Employees will know and understand that you track and analyze phone calls. This means they will be individually accountable for their actions. As a result, you get fewer personal calls made, call duration drops, and customer service improves.

#10: Reduction in bills: Typically, a call accounting system reduces a business's phone bill by 10-30%.[2] That gives you more money to invest in other areas of your business, so that you can increase sales.

---

2. Globaltech, http://www.globaltech-online.net/call_accounting/Calltrack%20CT2K%20Call%20Acctg%20System.pdf

CHAPTER 8

Call Recording Defined

"Your ability to communicate is an important tool in your pursuit of your goals, whether it is with your family, your co-workers or your clients and customers."

- LES BROWN

Call recording is a service in the telecommunication niche, that is used most often by customer service agencies to monitor calls from their clients. The management usually records these calls to check on employee performance.

It is also a great way for the business to handle customer complaints, and to track orders via the telephone. Call recording software is the most typical method of getting hold of call records for business purposes.

WHAT IS CALL RECORDING?

Unlike call accounting, call recording is used specifically to monitor, track and analyze the performance of its employees as they engage with customers telephonically. Call recording should be an important service for businesses looking for qualified performance improvements in the communications niche.

Call recording is often completely customizable, flexible and reliable – which means that you will not have to worry about trying to make it work in your business model. All great software should meet your individual needs once it has been customized by you.

Many recorded conversations can be used later for training purposes or for marketing materials. But these are not the only reasons why you would need to use call recording software. Risk management, quality control and dispute resolution are also made easier with recorded calls.

Most call recording systems are set to automatically record any and every call that comes through. This is especially true for medium and large-sized companies that are greatly concerned with customer relationship management and employee performance.

Call recording can also come in handy for your employees, and they can use it on demand if they feel like the conversation should be monitored. This can provide lawyers with evidence down the line, in case a customer has made false claims against your company.

With call recording, you should be able to steadily improve the performance of your telephonic workforce, while systematically ironing out the problems associated with control, accountability and work ethic.

KEY FEATURES OF RECORDING SYSTEMS

All quality call recording systems come with some key features that you will need in order to execute the previously mentioned strategies. Like many systems however, the kind that you get depends on your provider and the needs of your business. If you do not have a call center for example, you may not need as many features as a company that has one.

- **Excel export of logs:** Once the call logs have been compiled you can review them in PDF format, or export them into a workable Excel spread sheet.

- **Recording functionality:** You will get automatic recording options and on-demand recording options, for people in your company that need to use them.

- **Detailed search:** Search the call records by type, time, display information, comments, duration, user or one of the other main report categories.

- **Statistics:** Get all the stats you need to make innovative decisions regarding your employees and your business policies.

- **Security:** Permissions based user privileges and top notch security will keep these records from falling into the wrong hands.

- **Email records:** You can email your phone records directly from the call recording software in some circumstances.

- **Active IP recording:** Preserve call index information like incoming and outgoing phone numbers, time of call, search and replay, duration—when you activate the IP recording function.

- **Audio encryption:** Many audio files can be encrypted to prevent unauthorized playback of the calls.

- **Call muting:** Mute the call for a few moments while you check on something with another co-worker. Once mute is on, the person on the other side of the call cannot hear you – though the recording device is still on.

- **Contact visualizer:** A graphical user interface that shows captured interactions graphically—which is very useful when you need to quickly identify trends or patterns of behavior within your records.

- **Customer feedback app:** An app that automatically asks your caller a specific question and then records it for use in later marketing materials.

These are the main functions of call recording systems, though there are many more. A few of these include data compression, channel decoding, instant web replays, last call replays, multi-site recording and mobile phone recording. Take some time to fully understand what your call recording software can do.

KEY BENEFITS OF CALL RECORDING

Businesses work hard to understand how effective their sales teams are, and what kind of customer service they are providing their customers. Both of these extremely important factors can be improved with call recording. That is why knowing about the benefits of call recording is really the first step to improving your overall service offering.

- **Reduce corporate and personal liability:** The possibility of lawsuits is out there. One small miscommunication and you could end up paying a stranger millions. With call recording, there is evidence in place to defuse customer complaints before they turn into lawsuits.

- **Quality control:** Call recording helps ensure that what the customer wants, and what the call center agent or sales person has written down is the same thing. This prevents order and entry errors and guarantees that your product finds its way to your customer, based on their accurate details.

- **Training:** Few things are as effective as real life examples. When you are training a host of new sales people or call center agents, you will need these recorded examples to teach your new team how to handle themselves under pressure, while maintaining the integrity of the business.

- **Performance reviews:** How do you know if one call center agent or sales person is outshining the others? By reviewing their calls. The same goes for agents that are rude, or are not doing their job properly.

- **Improve customer satisfaction:** Call recording improves customer satisfaction, as managers are better equipped to train staff members and correct existing problems with dialogue and sales language.

- **Improved business security:** Security problems tend to reveal themselves when you record calls. It also enables you to reduce the amount of inappropriate calls being made on your work lines.

- **Reduction in personal calls:** If you suspect that your employees are calling using your dedicated phone lines, then call recording will find out for you. You can reduce these calls or eliminate them altogether.

- **PCI Compliance:** PCI compliance requires that you record phone conversations that handle credit card details with the utmost security. This way you prevent yourself from being liable for identity theft and stolen money.

These key benefits ensure that call recording remain an important part of the daily running of your business. If you have a call center or help line, you might want to consider modernizing to protect your interests, your customers and your employees.

QUALITY MANAGEMENT IN CALL RECORDING

In every business, quality management is paramount to success. It drives customer satisfaction and fuels repeat business. As you know, a company cannot survive without repeat business. If you can figure out a proven method of improving quality in your business—then you have a real opportunity. Call recording can do this for your company.

Quality management as it relates to call recording usually comes in a module or platform of some kind and ensures that your quality assurance and management staff have clear methods of evaluating your employee's levels of performance.

It also serves as a way to ensure adherence to set corporate procedures that were implemented by senior management. The ultimate goal is to maximize employee productivity, and call recording can go a long way in helping you achieve this milestone.

This can come in the form of call recording that works in conjunction with screen capture modules—so that you can rate calls while interacting with the audio. Many companies that offer call recording also offer different toolsets that you can use, that complement your call recording software.

Quality management at its core is about continuous improvement in quality and performance. Your call recording software should indicate where the weaknesses are in your company. By analyzing these insights after reviewing the calls, you will be able to ACT based on the collection of metrics that you have plotted from this software.

Unified communications that use quality management tools nearly always result in improvements for the user's business. Everything from employee performance, sales ratios, call durations, upselling and conferencing can be optimized based on the information that you will get from your call recording software.

Some quality management programs come with speech analytics and frameworks that you will be able to use to accurately turn metrics into actionable data. It is only when patterns have been identified and understood, that action can be taken to either improve the business, or to reduce errors in the business that happen on a daily basis.

Quality management can also come in a suite of tools including call recording, screen capture, assistant, call scoring and agent access. Many companies that offer these tools do it as an all-in-one solution for your call center needs.

RISK MANAGEMENT AND DISPUTE RESOLUTION

When you have thousands of people calling into your company, you can bet that a healthy amount of them will be complaints. Risk management is all about knowing how to handle those complaints in a manner that protects your brand and employee interests, while ensuring that the customer is satisfied with the outcome of the call.

Proof may often be necessary to quell or resolve disputes that arise between staff and customers. The staff member will have access to details about the customer to make the process easier. However, if something goes wrong and the customer becomes threatening—what happens?

The call recording software will have captured all the interactions between the staff member and the customer, including voice calls, desktop screen capture data and other details from the conversation. The trained staff member can then use these details as 'facts' to quickly and professionally defuse emotional situations.

Complaints can result from a breakdown in communication, or the unwillingness of a staff member to follow procedure. If the argument does escalate, you will be able to view the entire thing and take action to fix the situation. These detailed recordings and extras nearly always help the manager decide a few key things.

The first is where an issue came to light, and how it was dealt with. Then the manager can formulate a plan to fix the issue. They will check that the proper procedures were followed to get to the desired outcomes. During this process, they will uncover areas that need improvement.

Once this process is complete the manager should know what happened. If it was the fault of the customer, they will then proceed to placate the customer using logic and other tools in their sales arsenal, like discounts or free gifts. If the staff member was at fault, they will need to be spoken to, given a warning or outright fired – based on what happened.

Complaint management is a big part of risk management, and it needs to be handled according to company policy. Call recording goes a long way to ensuring that the right processes were acknowledged during the dispute. Once the manager understands what has really happened, they can take decisive action.

And this action is based on facts that were recorded by their quality management program and call recording system. Together, they work to guarantee the best customer service possible, and to protect the business from unscrupulous customers.

TEN REASONS WHY YOU NEED CALL RECORDING

If you are still wondering what purpose call recording could have in your business, then this is not the time to stop reading. The honest truth is that you SHOULD be recording your calls. It's part of an analytics system that has transformed small companies into corporate giants in just a few short years. Don't you want this kind of rapid growth?

#1: Training and development. There is nothing more valuable than teaching your staff how to manage calls, by using actual calls as examples. Use specific recorded calls as vital examples in your training programs.

#2: Security concerns. If your company offers advice or instruction over the phone, or you accept credit card payments, then you need call recording for security reasons.

#3: Risk management. Manage risk by emailing copies of calls to customers to defuse bad conversations or to avoid future issues.

#4: Improve sales performance. The whole reason why you are in business is to make money. Call recording helps you do this by improving the sales performance of your call center agents. They will benefit from improved techniques and strategies over time.

#5: Avoid disputes. Meeting with a customer on the phone means that you can record the conversation and avoid any disputes that arise. People are more likely to cause a scene on the phone, but it is also the best place to gather evidence to protect your brand.

#6: Test best practices. When you record calls you have a unique opportunity to try different marketing, sales or conversational strategies on your customers. Then you can record them and determine how well they worked at a later point.

#7: Affordable performance tool. Call recording may just be the most affordable quality performance improvement tool in existence. Why hire an expert, when you can gradually improve your business based on your own analytics?

#8: Monitor quality. Quality management is very important to businesses of all sizes. Make sure that your customers are getting the service they deserve every day!

#9: Quick responses to queries. When you have a customer call in that needs answers, call recording helps you get them. Respond in record time to customer questions.

#10: Proof of agreements. If someone has booked a spot, or made a reservation or a verbal agreement about something, record it with your call recording software. Hold them to it!

DIFFERENCES BETWEEN CALL ACCOUNTING AND CALL RECORDING

There are several main differences that separate call accounting and call recording. While many people get confused between the two, you will see why they are actually quite different. The basic functions of each system are very different.

Call accounting for example, is specifically used to keep track of dialed phone numbers and caller ID, with intent to use this information to sort out the budgetary needs of the company's telecommunication systems. Incoming calls are not recorded at all.

Call recording on the other hand, is used to specifically record ALL calls, incoming, outgoing and internal—with a focus on quality management, risk reduction and overall employee performance improvement.

As you can see, though they work along similar lines, they both have very different functions in the business. Here are a few more differences to help you better understand why the two technologies stand so close together, yet operate independently of each other.

- Call accounting gathers analytics based on internal usage. These analytics are meant to be used to monitor telephonic expenses and reduce costs.

- Call recording gathers data from conversations based on internal, incoming and outgoing phone calls. This data is used for a wide variety of purposes, mainly to improve a host of business processes.

- Call accounting is used by the management team to monitor financial expenditure.

- Call recording is used by managers and staff, to regulate business processes, improve performance and resolve disputes with customers.

- Call accounting takes care of the money-side of telecommunications, while call recording takes care of the people side of telecommunications.

Both of these technologies should be used in unified communications to help the business improve their communication capabilities across the board. If your business can accurately monitor your telephony expenses AND use call recording analytics to greatly improve old models in the business, then rapid growth will occur.

When you consistently work on streamlining vital processes such as finance and human performance in telecommunications, your business will succeed in establishing itself as a quality brand. With satisfied customers, motivated staff and strict budgets you can take your business into the twenty-first century with these unified communication elements.

CHAPTER 9

Call Center Defined

"The Internet is the most important single development in the history of human communication since the invention of call waiting."

- DAVE BARRY

A call center is a physical location used by businesses to handle customer or supplier calls. Some of the services are automated, because they include features like voicemail and automated attendant. This service will direct you to the first available call center agent, when they are finished with their last call.

Customers use call centers when they want to ask a particular company a question, or provide them with some feedback on a product, or to register a complaint. Call centers are usually fairly large and have the capacity to accept high volumes of calls, all at once.

WHAT IS A CALL CENTER?

A call center might conjure images of sales people phoning you for no reason, but these are telemarketers, not call center agents. In fact, an organizations call center is an extremely important location, where calls are screened, forwarded to key people or handled on site.

Nearly all of these calls are recorded for reasons we discussed in the previous chapter. Many technology-based companies have call centers that act as 'helpdesks' for their customers. But many other businesses use them—such as mail order catalog organizations, large product websites and brands that ship things overseas.

Many companies are choosing to outsource their call center requirements to overseas businesses, because they are cheaper. While call centers have a poor reputation in the media – most of them work well and provide a valuable service to the customer.

These call centers are actually very complex operations with a lot to manage all at once. They function as the voice of a business, but have to deal with processes, technologies, people, customers, facilities and everything else a regular business has to deal with.

Call centers are run by highly efficient and results driven managers that use unified communications to positively influence their workforce. Customer demand for direct business interaction is on the rise due to new platforms like social media, which means call centers are bigger than ever before.

AUTOMATED CALL DISTRIBUTION - ACD

Automated call distribution is used to recognize, answer and route calls for your business. These systems (also called ACDs) are scalable, and they can be very small or incredibly large and used in a call center.

The larger systems can handle hundreds, even thousands of lines at once. When the automated call distribution system gets an incoming call, it looks around for instructions on how to handle that call. These instructions are preprogramed into the system before it works, and they can be altered as needed.

Then the ACD routes the call to a call center agent, or an automated recorded message. Alternatively, the caller is placed on hold until an agent is free. In this way a call center is able to run without the added stress of routing the thousands of incoming calls. It is all done on an automated level by the ACD.

The ACD can also be programmed to respond to a number of different handling instructions, according to the number that the caller selects. In this way the ACD sorts out what the caller wants, and where the most logical place is to send them.

These ACDs provide some kind of automatic customer caller identification (ACIS), like those provided by DID or direct inward dialing, DNIS or dialed number identification service or ANI, automatic number identification.

These are robust systems that allow the call centers to distribute queued calls in an organized, linear manner. With ACD the call center manager can instantly manage all of the crucial configurations of every call queue on the system. It also tracks queue statistics that can later be integrated into a quality analytics program.

With these special queue analytics and stats, the manager will be able to see how the agents or queues have performed on any given day. A few key features include:

- Custom greetings can be recorded for every individual queue
- Visual indicators on the hardware show the status of each queue line
- Direct calls to managers, co-workers or specific call agents
- The function keys are programmable which means they can be set to light up to represent different call queues
- Individual phones can be set up to respond differently to call conditions, including different types of ring tone.

ENHANCED EFFICIENCY

Call centers come with enhanced efficiency these days. Gone are the days of calling one of these centers, and having to spend 4 hours in a queue. Then when you reach a call center agent, they ask you the same personal details that they ask you every single time you call. It does not make for a very good customer experience.

That is why technology evolves, and new efficient programs are added to support call center technology. Screen pops were designed to be the bridge that call centers were missing. With this bridge they can finally give clients the customer service they deserve.

Screen pop technology was built around Computer Telephony Integration, and is a proven solution to this common problem. So you won't have to expose your customers to constant redirects in a call center, and 3 or 4 repeats of the same information.

A screen pop appears after the customer has identified themselves. When they are transferred the information is sent along with them and pops onto the new call center agent's screen. They address you by name and ask if they can help. Now that is customer service!

Screen pops are not new technology by any means, but they have – until recently – been out of reach for small and medium sized businesses. These days however, there are screen pop solutions that are highly affordable, and they have been proven to work very efficiently for these entry-level businesses with limited budgets.

Screen pops help you reduce the average length of a customer call, which in the long term can save you hundreds, even thousands of dollars. Many business owners that run call centers call these highly efficient screen pops 'mission critical.'

The real benefit however, is the increased customer satisfaction from the improved experience of dealing with call center staff that know who you are. Call centers that use enhanced efficiencies like screen pops do a world of good for their brand.

Even the agent's performance improves from using this enhanced technology. You can test it for yourself. When experience improves for the customer, performance improves for the agent. That's just good business!

CALL CENTER INTEGRATION

How would you go about integrating a call center into your business telephone systems? First, you would contact a trusted provider, the same company that handles unified communications and complete A-Z telephone services.

A call center integration service helps you by giving you the power to align your daily operations management with the priorities of your business. You will be able to increase customer loyalty by resolving issues as they happen, on the first call. You will gain new sources of revenue by taking advantage of cross and upselling techniques.

Plus, you will be able to control the costs of your business by implementing new, innovative techniques over time, and eliminating or reducing practices that are out-dated or don't work anymore. There are very few ways to use quality analytics in business, collecting them from your call center is one of them.

Find a company that will improve your core contact center procedures. They will help you set up a scalable call center that has everything you need to begin running your business properly. It is only when you sort out the communication in your business, that real progress will occur. With real feedback, there is nothing you cannot achieve!

From operations integration to call center planning, design and implementation, what you need is a company that knows you are serious about growth and income expansion. These professionals will guide you through the call center integration process by recommending the right hardware and software systems to use, as well as the best analytics data available.

You will then be able to setup your very own call center, which will pay for itself over and over again. With the insights that you receive from these customer calls, your income is bound to rapidly improve in the coming year.

A call center integration specialist is someone that will assess the needs of your business, and actually build you a quality call center infrastructure from the ground up. Once this is done, it just needs to be staffed and managed. If you can operate a successful call center, it will be the fuel that carries your business into the big leagues.

HOW CAN A CALL CENTER HELP MY BUSINESS?

Establishing a call center for your business may be the best thing you have done for it this year. There is no avoiding the facts, and that includes how customers want immediate responses and results—and they want them now. If your business cannot give them this level of personal service, then they will move on to a business that can.

With small businesses, a call center may seem like a preposterous idea. You do not have the time or money to manage something on that scale! But what you don't know is that you can develop a small call center that acts as a centralized place for all of your customer queries and complaints.

These small call centers are relatively inexpensive to setup, and need a unified communications infrastructure. They help save employees time, and they increase profit because they improve productivity.

A restaurant owner for example, might ask what the point of a call center is for his business. The phone on the other hand could be ringing all day, every day – for take away orders. The owner spends all day on the phone and misses a lot of calls. If this was handled better, the owner could be free to run the restaurant, while automated services and a small call center deals with the influx of orders.

Better yet, the customer feedback from the orders would result in better food, more customers and greater exposure for his restaurant over time. As you can see, this restaurant owner can really benefit from a small call center.

If you take the lessons you have learned in this book and apply it to your business, you can create an advanced, affordable call center that will get your more business, and skyrocket your sales. As long as you have telephones, you have the opportunity to market, and manage customer expectations.

An alternative may be to outsource your call center needs to a call center company. They will handle all of your calls, while you get busy doing what you do best—business growth! Every individual company is different. You are the only one that can determine whether you need a call center or not, but take heed.

As customers become more and more selective they will prefer to deal with businesses that have taken the time to be accessible. Accessibility and availability are the two features that will dictate how large your business is able to grow in the future. With a call center, there is no growth limit. Your customers will be happy, and your business will be better.

CHAPTER 10

Optional Equipment

"The utility model of computing - computing resources delivered over the network in much the same way that electricity or telephone service reaches our homes and offices today - makes more sense than ever."

- SCOTT MCNEALY

When choosing optional telephone equipment for your business, you need to think about usage, need and the benefits that come from investing in additional equipment. Sometimes it is a lot more affordable to lease equipment on a temporary basis, or you could buy it outright—depending on your immediate needs.

Optional equipment includes headsets, softphones, wireless solutions and paging systems. You will need to review each of these to understand how they can positively impact your business in the short and long term.

HOW CAN OPTIONAL EQUIPMENT HELP YOUR BUSINESS

Any business that wants to sort out their communications structure, needs to have insight into the smaller, less important 'functions' that make great business telephone systems work. The optional equipment items on offer can enhance the user experience quite a lot. Here is how investing in these optional extras can help your business.

- **Time saving:** Using better equipment means that your staff will always be available to take calls, even when they are busy doing something else. At the click of a button, calls can be taken and processed.

- **Money saving:** Instead of buying 30 new telephones, why not invest in some headsets? They save you money and they make accessing telephone calls easier for call center agents.

- **Improves speed:** Being able to handle calls quickly and easily is part of giving your customers the quality of service that they deserve. When your staff speed improves, so does your income and customer satisfaction.

- **Increases productivity:** If you can guarantee that productivity will benefit from optional equipment, then you need it. Anything that will help your staff do a good job is worth looking into, especially in the field of communications.

- **Improves internal cooperation:** When your staff find it easier to work together, they do a better job and everyone wins. Optional equipment can result in improved connections between employees, which always boosts business.

There are dozens of other ways optional equipment can help you improve the communication structure of your business. To fully understand which core components can achieve these improvements, we outline the main equipment items as you read on.

HEADSETS

What are the main differences between using a telephone or a headset? There are lots of them! A headset may be an optional extra, but business owners of call centers know how valuable these slight additions can be to the daily running of their company.

- *Headsets improve productivity and freedom.* They free up both of your employee hands so that they can type on the computer, search for important documents or find answers to client questions at high speed.

- *Headsets give you freedom of movement.* With a handset lifter you can be completely away from your desk and still handling calls. This means you can walk and talk – answer and end calls on your own terms. Less voicemail means more productivity.

- *Headsets that are wireless give staff the power to leave their desks.* It can be difficult being trapped at a desk all day. When you give your staff the flexibility to move around, they can go as far as 300 feet away from their desks, while working.

- *Less background noise.* Telephones tend to pick up a lot of noise interference from ambient locations. When you use headsets, 75% of background sounds can be filtered out, which means your customers can only hear your conversation.

- *Constant call and voice quality.* When you wear a headset the microphone stays in the same place and position. This means that the volume and quality of your voice does not distort during the call, as it does on your average phone.

- *Headsets improve focus.* When your employees wear headsets it forces them to be completely focused on their callers, and not on what their co-workers are doing. This means that they will get through more calls a day, thanks to improved attention-span.

- *Reduces back pain and neck pain.* In call center jobs, or jobs that require a lot of telephone usage, it is not uncommon for the employee to suffer from bad posture, back and neck pain as a result of answering phones all day. Headsets negate this side-effect, which is why doctors and chiropractors recommend them.

Headsets are always sold as optional extras, when in fact they are more efficient and easy-to-use than telephones. If you have a receptionist or a cell center, then you will want to think about replacing your phones with headsets. They are better for your staff, they improve customer service and they increase productivity.

WIRELESS SOLUTIONS

Above all, employees need flexibility in the workplace when dealing with customers on the phone, and co-workers in the office. In today's modern workspace it is rare that employees spend all day at their desks anymore—they are nearly always on the go.

Whether they are going to a meeting, seeing customers, getting coffee or hosting an internal presentation – they need to be connected with their apps and incoming client calls. Wired phones are great, but are inflexible. That is why it is important that you consider wireless telephony solutions when you are reviewing optional extras for your business.

- *Cost-effective and reliable:* Mobile communication is affordable, and it is a very reliable way of ensuring that your employees are permanently connected, even when they are not at their desks. Things move fast in the business world, and you have to stay connected to keep up these days.

- *Secure wireless solutions:* Keeping your business conversations private and confidential is paramount to your business. That is why you need to find a wireless solution that offers security against unscrupulous people.

- *Increased productivity:* Employees that are mobile and equipped with wireless handsets are easily more productive, as they do not lose work hours when they leave their desks. These staff members are also happier with their jobs, and get along better with management and their fellow co-workers.

- *Improved customer satisfaction:* When customers get through to a real person every time, it does wonders for your business reputation. Plus the fact that your employee always takes the call, means that problems are solved faster and a rise in customer satisfaction occurs.

- *Focus on quality, durability and functionality:* Most wireless handsets and solutions are very high quality and they last a long time for an item that it used all day, every day. Best of all, they are highly functional – linking your employee with a multitude of functions that they can perform for improved customer service.

Many of these wireless handsets are lightweight, they have a special ergonomic design for comfort and they are resistant to shock, dust and liquid damage. Your employees can roam around your offices, getting work done—as they attend to the needs of your callers.

PAGING SYSTEMS

Even though pagers and paging systems are considered telephony optional equipment items, they can be indispensible to the right business model. If reliability and coverage matter more than anything else, you will want to consider a paging system in your unified communications infrastructure.

- *The most reliable system.* There is a reason why doctors use pagers and that is because they are 99.9% reliable. It uses simulcasting (radio signals) to send and receive messages, which means it works even if one or more transmitters have stopped working.

- *Paging systems are low cost.* High quality pagers are low cost, when you compare them to other telephony technologies. Operators charge a fixed cost for the system, which means you do not have to pay per message. Send one million messages a day for the agreed upon set rate.

- *Very wide coverage:* Where other messaging technologies fail, pagers succeed. They tend to work in areas other systems do not, like tunnels, underground, on airplanes and in hospitals.

- *No network congestion.* Paging systems are ideal for emergency situations because they do not suffer from network congestion. When a major crisis is happening, other networks slow down, and messages do not get through. If your business relies on messages getting through, then pagers could work for you.

- *Paging systems can be integrated.* Paging systems do not have to stand on their own, they can be integrated with other systems to become part of your unified communications infrastructure. Link them with IT systems, production lines, smoke alarms or apps.

- *Company-wide calling.* With a paging system, you can call everyone at once – something you cannot do with other kinds of telephone systems. This is part of the reason why it is superb for emergency services and other companies that deal with their employees as a whole.

- *Speedy delivery.* You can rely on a pager message being delivered in 5 seconds and no more than 30 seconds, every time. There is literally no delay on the message when it is sent through. It gets sent, and it arrives!

- *Paging systems are discreet.* Great for top management teams, these paging systems are discreet and excellent for on-the-go employees. They are small and compact, and they connect you to everyone else.

COMMERCIAL ON HOLD PRODUCTION

Commercial on hold production is fairly unknown to small and medium businesses, but it is a great opportunity to capitalize on some free time. When a customer calls in, and the call center agent has to put them on hold, usually there is some bad music or tone playing in the background. Why not use this time to customize your on hold message and make it count?

A 'commercial on hold' is a customized advert that plays when your customers are waiting to be served by your call center. They usually promote, explain, inform or upsell – which helps the customer make informed purchasing decisions while they wait.

- *Customizable messages*: Record anything you think your customers need to hear, to improve customer service, sales or general information sharing.

- *Productions from scratch:* The best on hold productions are made by all-in-one business telephone systems suppliers. They will create a fresh, creative commercial for you that will prompt action in your customers.

- *Auto-play function:* Whenever a customer is put on hold, they will be exposed to these excellent commercials, which act as audio advertising for your company.

- *Added music*: Just in case you want to take a break from advertising, many commercial on hold systems come with music built-in.

- *Automated attendant*: To get your callers to the right advert or call center agent, the auto-attendant will first ask them to choose their call path.

- *Optimizes exposure*: In business, you need to leverage every moment to get the maximum amount of exposure possible. When your customers hear your audio commercial, they are more likely to buy from you. It is that simple!

- *Using your time wisely*: It has been proven that over 70% of all business phone callers are put on hold – sometimes for over 10 minutes. This is a great opportunity to entertain them with a creative audio commercial.

- *Increased call retention*: Another great side-effect of entertaining commercials, is that callers tend to stay on the phone longer. They will listen to your commercial longer than any music, so that your call center agent can eventually serve them.

- *On hold message preference*: It has been proven that people prefer on hold messages to music or silence. Cellular Marketing Magazine reported that over 85%[3] of phone callers prefer messages to other forms of on hold distractions.

3. Customized On Hold Messages!, http://www.commercialsonhold.com/

CHAPTER 11

Your Carrier Options

"I really see low-fare carriers, quality low-fare carriers anyway, continuing to become more and more popular."

- DAVID NEELEMAN

One of the key decisions that you will be required to make in business, is which telephone carrier you decide to go with. There are many, many telephone options out there, and they are definitely not all equal.

Choosing a telephone provider will require knowledge on the main telephone services that each provider offers. Take a look at these services and decide which one is best for your business, in the short and long term.

POTS LINES

POTs lines also stand for Plain Old Telephone Service, and mostly refer to the basic telephone service that is used widely in homes and businesses across the globe. A POTs network is also known as a Public Switched Telephone Network or (PSTN).

POTs lines are still the most widely used, of all the telephone services currently in use in America. These analog phone lines are individual lines that can be used by any business for daily phone calls.

POTs lines work even if there is a power outage as they have their own power supply. This is why it is common for phones to work, when everything else has stopped working in the office. Each line on this system has its own designated telephone number. If you need another phone, you will have to get another line – that is just how it works.

A little outdated for modern businesses that use a lot of telephony equipment, POTs can become expensive to operate due to the taxes and surcharges involved. While one business phone line can cost around $40.00 per month, having dozens of these lines can add up, resulting in large phone bills that your average business cannot handle.

POTs lines are however the most economical telephone service for small businesses that do not require a lot of telephone usage. Keep in mind that it is not unusual to be billed by third parties for a service that you 'added' to your line over time.

Installation of these multiple lines is very costly however, and it is not a good telephone service for businesses that need quality, high volume business telephone systems. Initially, you will think that it will not cost much, but the bills you receive for installation will shock you!

SIP – SESSION INITIATION PROTOCOL

Another top quality telephone service is SIP, or Session Initiation Protocol. This particular type of service is used for controlling multimedia communication 'sessions' over the internet – like voice or video chat calls. In other words, it is a signaling communications protocol that works in conjunction with Internet Protocol networks.

SIP can be used in your company by your employees to communicate across LAN, WAN or Internet networks. They connect incoming and outgoing calls over the Internet, which is far less expensive than connecting calls over an ISDN line, which is a fixed telephone connection.

When you use an advanced service like SIP, you will have access to numerous apps, including video conferencing, instant messaging, document sharing, or general audio chatting. When all of these networked apps work together, we call it unified communications—because they are all monitored or governed under one set system or service.

There are so many new and exciting benefits to using a service like SIP over other older forms of telephony services. Money savings are the most popular, because Internet calls between certain programs are free, and cost your business nothing.

In addition, calls to cell phones or other landlines via an integrated software program are much cheaper than the rates you will pay for calling on the average phone line. SIP is highly flexible and comes with a wide variety of additional services and apps that can be customized and shaped into the ideal unified communications infrastructure for your business.

SIP trunking negates the old method of having to connect with the telephone company in order to make a call. Instead, you do not need a dedicated interface like a phone line—you can simply connect via your broadband Internet connection.

The good news is that Internet lines can carry far more calls than a phone line, and congestion is rare. Using SIP you can add as many external phones as you need without having to add any physical lines into your business. This saves you time, money and hassle – especially when your business is growing rapidly.

Enjoy the increased efficiency of SIP as the single connection easily allocates bandwidth to your voice and data needs during the day. Your customer service will improve thanks to the many integrated business apps that help you serve your clients efficiently. If your business needs geographic independence, simplicity, growth and reduced costs, SIP is for you.

PRI – PRIMARY RATE INTERFACE

The PRI, or Primary Rate Interface is a telephone service level that is part of the ISDN specification for carrying many data and voice transmissions between a user and their network. Put simply, PRI is one of two levels of service in ISDN. The other is BRI, or Basic Rate Interface, but this is usually used by small businesses.

PRI is used by larger companies, and the service is a proven network technology that works with your office phone system to help your company manage mass communication. DID, or Direct Inward Dialing means that all employees have their own extensions, and can have calls routed there. In this instance there is no need for an automated attendant.

There are 23 voice channels in PRI and a signaling channel in each of these circuits. This system gives you all the advantages of advanced digital calling, and is used most often as the standard telecommunications service to offices.

This PRI service includes a number of B-channels and D-channels. B-channels carry data, voice, apps and other services, while D-channels carry control and signaling information. This service is typically used by medium to large sized companies with digital PBX systems, so that they have digital access to the PSTN, or Public Switched Telephone Network.

All 23 B-channels can be reassigned and are very flexible, which means your business can use a variety of media as you need it. Video conferencing for example, is one benefit of having a PRI service as a large business.

This local exchange access service provides high speed direct digital connection to the phone line – which means that even though you are using a traditional phone line, you still have access to the latest digital telephone services.

- Multi-channel service that supports different services on different channels

- Local exchange access for high speed digital connections
- Digital service integration, allowing you access to packet switch data in a single loop

If your medium or large business has a traditional PBX system, but you need to upgrade to allow for digital app and service use, then you may want to consider a PRI service. If you are a small business, you can consider the BRI service, which is the same thing only it scales to suit the needs of small businesses better than PRI services.

Dynamic Integrated PRI

There are two types of PRI (Primary Rate Interface) that you can get. These are static or dynamic PRI services. A static integrated PRI has pre-determined data and voice channels. They are completely fixed. A dynamic PRI however, allows for 23 channels to be used for data and voice services, on an 'as-required' basis. These are flexible!

This is how dynamic PRI works. If you are only using two channels for voice calling, the dynamic allocation function will automatically place the other 21 channels in 'data mode' which increases your overall data speed. As soon as a voice call is in session, it will assign a B channel for that call, and the speed will reduce slightly.

The difference is that voice traffic has priority over other forms of data connection. All channels carry data to increase bandwidth, but voice traffic will always get first priority which is very beneficial to a large or medium-sized business.

When you get dynamic integrated PRI, you eliminate the need for separate voice and data services from multiple vendors, which reduces your system expenses. There will always be a certain number of channels specially reserved for voice, while the rest of the channels deal with data. The moment a phone call is disconnected, the data speed increases.

Best of all, dynamic integrated PRI usually comes with equipment included, so you do not have to spend extra on additions for your business. There is usually DID capability, which allows incoming calls to be served by a PBX without the need for an automated attendant.

It also uses call by call allocation, which enables all 23 channels to be used for incoming and outgoing calls - or data transmission. The benefits of using this kind of integrated system are clear—you get dedicated trunk groups that can be individually configured, and enhanced 911 features which means that the DID number is transmitted over the D-channel, so that the callers number appears on the recipients phone display.

Dynamic integrated PRI is also highly efficient, as it consolidates traffic from a PBX, which reduces equipment costs. Most of all, this sort of system is affordable and flexible – so you can handle more calls on a single digital connection.

DID - DIRECT INWARD DIAL

DID, or Direct Inward Dialing, is also called DDI or Direct Dial-In, and it's a feature that telephone service providers offer their customers for use with their PBX systems. One or more trunk lines are provided so that the customer can connect to the PBX, and a number of telephone numbers are allocated to this line.

The goal of this service is to allow every individual employee to have their own personal phone number—only you do not have to install separate phone lines to do it. Telephone traffic is split up and managed in a more effective manner.

To get a DID service, you first have to have an ISDN (digital line) installed. Then the telephone company will provide you with a block of telephone numbers that run through the companies PBX system. This system automatically routes callers to the correct people.

If you have a company of 50 people, and everyone needs their own phone number—you can rent these numbers from your local exchange carrier. These calls would operate over 8 physical telephone lines (trunk lines) and would allow for 8 simultaneous calls. The PBX automatically switches a call for the given phone number to the right workstation.

DID systems are great for voice mail, fax and live voice connections. They are better than your standard PBX service, as DID saves you the added cost of a switchboard operator. Calls also tend to be routed faster, and customers benefit by not interacting with a computerized voice message.

If your company needs a PBX system, but also has additional needs for separate, independent phone numbers for your staff, you should consider getting the DID service from your phone exchange. Like many added phone services, these will actually help your business save money over time, by eliminating unnecessary employee positions and reducing overall call durations and costs.

DID or Direct Inward Dial is currently being used by many successful businesses all over the globe. DID is widespread, and has coverage in all major rate centers. There are flexible routing options available which make it easier to have local telephone number presence in many areas. These calls are then directed to a central location for easy management.

WHAT YOU NEED TO KNOW ABOUT CARRIER PROVIDERS

There are dozens of carrier providers in the country, but you need to understand how they work, and which one would suit your company best. A carrier provider will give you the telephone services you need to improve your unified communications structure.

A carrier is an owner of a cable or wireless facility, which is used to supply 'carriage' services like telephony or internet services to the public. They are also called local exchange carriers or just local telephone companies.

These local exchange carriers carry telephone calls and other communication services that use telephone lines.

- There are two different kinds of carriers—the long distance carrier and the local telephone service provider. Other kinds of carrier do exist, but they are known as business phone service providers.
- Some carrier providers are more advanced than others. They should offer you dozens of features, like auto-attendants, network monitoring, call park and anonymous call rejection.
- Some telephone companies may try to upsell their non-Internet based technology, because it makes them more money. You need to understand that it is important your business has digital access, and can work with different media online.
- Sometimes all you need is a computer, some software, speakers and a headset. Anything else that is recommended to you could be excess. You will need a stable Internet connection, or a business telephone system that works in conjunction with the Internet.
- You will have a choice of hardware and software, always go with the package that best suits your needs, not the one that the sales person suggests. Review all of your options before choosing the unified communications structure you need.
- A lot of small independent 'companies' have sprung up offering business telephone services. This is because VoIP allows anyone to be an Internet Telephone Service Provider. Never use these small companies, as their servers are unreliable and inefficient.

- Chat with your provider about porting and moving your phone number if you need to. This is never a quick thing to do, so it will tell you a lot about that provider.

- You get what you pay for when you choose a telephone provider. If you go with a small company, expect erratic services and the possibility of losing all your business phone numbers. Bigger is better in this niche, and so is reputation.

CHOOSING THE RIGHT CARRIER

How do you decide which carrier is the right one for your company to use? There are a number of key indicators that you will have to run through before deciding on your permanent business phone service provider. Here are some helpful checks to keep you looking at the right companies.

- *Customer service:* One of the main issues that businesses have when they choose their business phone service provider, is that the staff aren't helpful enough to support your business implementation and phone management. Always test a company's customer service lines to see if they are any good.

- *Compare features:* Your business has specific needs. If you explain your needs to the carrier and they cannot suggest a unified communications infrastructure that works, perhaps they are not any use to you. Your phone company must give you the features you require for success. Compare the main services of each company.

- *Compare pricing:* Pricing can tell you a lot about a carrier company. Going with a slightly less expensive carrier can be disastrous, if they are not well-established or widely-used. Only compare the pricing structures of highly reputable and long-standing carrier companies.

- *Network speed and quality:* If you are going to be using VoIP services, make sure that the actual NETWORK functions as advertised. A quality network will not drop your calls, have low quality sound or give you less data than laid out in your agreement. Wireless speeds and network capabilities need to be tested and checked.

- *Traffic shaping:* A common problem with carriers is that they sell you services and then shape your line speeds. If they shape your lines too much, it will have terrible consequences for your voice calls and data transfers.

- *Check on the hardware:* Your carrier should provide you with a wide range of hardware and software choices. Whether they are analog or digital, you will need to assess which company has the best software and hardware overall for your business.

Choosing a reliable phone service provider is really about comparisons, testing and ensuring that they live up to their side of the bargain. You might find that after four months of using a 'great' service, suddenly your line speeds drop, or your call quality deteriorates. If this happens, get on the phone and make sure they know that you are not getting the services that you are paying for.

CHAPTER 12

Pre-Owned Telephone Systems

"To effectively communicate, we must realize that we are all different in the way we perceive the world and use this understanding as a guide to our communication with others."

\- ANTHONY ROBBINS

If you are looking to establish a business telephone system that does not cost you your entire capital outlay, you may benefit from purchasing a pre-owned system. When you need a lot of hardware, and want to drop the prices on the equipment, pre-owned systems can go a long way in ensuring that you get what you need, at the right price.

These pre-owned systems can also come in handy if you need to replace a portion of your existing system - but you do not want something completely new. Buying second hand phone parts is valuable to the business owner that does not want to overspend.

EVALUATING PRE-OWNED EQUIPMENT

When costs are high and so are your business telephone needs, sometimes choosing a second hand system is exactly what you need to do. But this usually results in additional problems. How do you know the system is still performing at its best? This is how you need to evaluate your pre-owned equipment.

- Companies that offer pre-owned equipment should refurbish, review, clean-up and intimately assess the state of the system before selling it. To find out if they do this, speak to the sales people and ask them about their evaluation process.

- You can always invite a business telephone systems engineer to check out a second-hand system that you are interested in. They know more about it than you do, and will be able to see immediately if anything is out of place or broken.

- Always look for a guarantee on the business telephone system. Guarantees are great because if something does break, the company is liable and they have to replace the broken part. There are usually twelve-month manufacturer warranties as well.

- Short of getting an expert to help you, you can also test the system out for yourself by installing it and seeing if everything works efficiently. Tell your employees to make a note of what does not work. Then hand this list to the service provider.

If that fails, try educating yourself on a more advanced level about business telephone systems. This book does a lot to educate you, but you can also use Internet resources to check-up on certain performance features as you go.

REFURBISHED TELEPHONE SYSTEMS

A refurbished telephone system is a system that was once bought and used by another company. Since then, the company has upgraded, closed, or perhaps moved onto a different type of telephone service altogether.

The term refurbished is used to describe an electronic device that has been restored to its original functionality. They call these systems 'like-new' because they operate exactly the same way that new systems do, just at a fraction of the cost.

It is like buying a really good pre-owned car. As long as you know how to evaluate them (with expert help) then you can save a lot of money by getting one over a newer system. Most refurbished phone systems come with a 12 month warranty, because they have been fixed, tested, cleaned and reboxed by the manufacturer.

These companies will have a very strict policy that outlines what they will buy for refurbishment or not. They will not pay for a system that they cannot resell, but this does not mean problems don't persist. Often with electronics small glitches can go unnoticed, so you will need to look out for those.

On the whole these manufacturers are very good at buying quality pre-owned phone systems and then fixing them for resale. It is a way that they can continue offering their services to a wider range of people at a reduced rate.

The truth is that many of these business telephone systems are more than adequate. Even after years of use, as long as the hardware is functional and the software is clean, it will meet your needs.

Business owners are naturally suspicious of refurbished business telephone systems, and problems can arise. But most second-hand customers are perfectly happy with their purchase, and go on to use that system for several more years.

For new businesses that need to control spending on infrastructure, second hand telephone systems is an ideal way to get what you need, without having to fork over thousands of dollars. For a medium-sized call center for example, you can save a large amount of money if you just choose to buy second hand.

PROS AND CONS OF BUYING EQUIPMENT ON THE INTERNET

As in any electronic niche, there are pros and cons involved with buying business telephone equipment online. The online marketplace is home to more dodgy sellers and 'once-off' refurbishers than you can imagine. But it is also the playground of legitimate businesses that want to give you and your business a great deal.

These companies make it easy for you to see how credible they are, so that you do not get confused between them and the 'dodgy' businesses that will take your money and run. Here are some pros and cons of buying telephonic equipment online.

The Pros

- Buying online often means greater choice, between brands and between systems. If you want to find something that ideally suits your business, there is no greater arena than the Internet.

- Buying business telephone equipment online can be a lot cheaper, if you are also buying from a local source. Sellers purposefully run promotions on their sales websites to trump up more business—and you get the benefits! In addition, you can buy second hand equipment direct from the original manufacturer.

- When you buy equipment online you are able to take advantage of the many comparison websites online that lead to better purchasing decisions. Things like user reviews, social proof and customer feedback is also available for you.

- If you need something fast, the Internet has it. If all local shops have failed you, there is bound to be one online that can help you find a phone part or whatever you need. Then you just need to order it, get overnight delivery and presto!

The Cons

- There are unscrupulous businesses and business people that are just out to make a quick buck off you. They set up fake or poorly run businesses and then sell you tragically bad equipment. These same people are never reachable after you realize that you have been caught out.

- If a store goes bankrupt, as many online stores do—it makes the warranty for your equipment null and void. This means that even if you did get a great refurbished product, if it does not last—there is nothing you can do about it.

- You cannot inspect the item before you buy it. There are stock standard pictures on these websites and some are downright false. What you pay for and what you get can be very far apart if you are not careful.

WHAT CAN YOU EXPECT?

It is hard to say what you should expect with refurbished equipment. It really depends on the company that you choose to use. If the company is real and customer-centric they will ensure that your refurbished phone system works like a charm.

- Expect a phone system that works as advertised. That means there should be no discernable different between the new model and the second hand model, once it has been refurbished.

- Expect identical models, not models that are 10 years old and masquerade as the same thing. If the image is the same, that is what you should be getting. Check that the hardware and software is identical to the new original.

- When a used phone system is installed by your business phone provider, they will make sure that all of your staff are trained to use it, and that it is set up for use from the very first day.

- Most manufacturers or business telephone companies provide free or paid lifetime support for a phone system that you buy from them, regardless of whether it was new or second hand.

- Expect a one-year warranty on any and all refurbished business telephone systems, including software and hardware.

- Expect good refurbished systems to be from the large brands like Alcatel, Samsung and Panasonic – which guarantees their durability and quality.

- Expect qualified, certified technicians to come to your offices and install the refurbished telephone system, just like they would with a new one.

- Expect support regarding replacement parts or spare parts if you need any by the original supplier or manufacturer.

- Expect to be sold a refurbished system that meets your growth needs and does not contain any out-dated parts or hardware that you cannot get anymore.

In other words, you should expect the same quality service from your business telephone service provider, only they will be installing a more affordable, second hand system—instead of a brand new, out-of-the box system that will cost you a lot more.

For start-ups this could not be a more attractive way of ensuring that your business has the communications infrastructure it needs without overspending on the initial set up. It is exactly why more and more small companies are beginning with refurbished telephone systems.

CHAPTER 13

Signs You Need to Upgrade

"The only thing nicer than a phone that didn't ring all the time (or indeed at all) was six phones that didn't ring all the time (or indeed at all)."

– DOUGLAS ADAMS

One of the key problems areas in business is the unwillingness to upgrade, and this is definitely true of business telephone systems. Businesses implement systems and strategies and then 'forget' to upgrade.

This forgetfulness is usually due to the fact that setting up a new system can be costly and incredibly time consuming. However, it is exactly what you should be doing, every two years. Here is how to tell if your business telephone system needs to be upgraded.

YOUR TELEPHONE SYSTEM IS TECHNICALLY OBSOLETE

If your business telephone is visibly or technically obsolete, then you need to seriously consider upgrading it. There are real dangers to operating a business telephone system that has been discontinued by the manufacturer.

Unified communications is about connectivity, convergence and integration—it is about leveraging powerful technology to improve your business in many ways. More and more, businesses are realizing that their communications systems are not living up to the functionality or flexibility of this new kind of technology infrastructure.

You need to review the compatibility and interoperability of your phone system in relation to any new tech or networking equipment that you have purchased. If it is outstandingly incompatible—then it is obsolete.

There are many business telephone systems that have been discontinued by manufacturers, which is a ticking time bomb for your company. The time is coming when your parts will run out; your service men will stop using the tech and incompatibility becomes so rife that you will be forced to buy a new system.

Of course at this point the damage has been done already. You have probably bought bits and pieces of other systems in an attempt to keep your old one going. Your staff has become used to this out-dated technology and will struggle with any changes.

The bottom line is that if you run your business on obsolete technology, eventually there will be no-one to fix it, and it will break. When that happens your company will be without a communications system, and all you'll have left is a business full of people that have no idea how to use more progressive systems. It is a massive financial loss waiting to happen.

EXCESSIVE SERVICE

If, like many businesses in transition, you suffer from giving your customers 'excessive service' you need to sort this out before you expand. Your communications structure is key to your customer relationship management, but it can get out of hand if you do not have the right tools on your side.

Yes, we live in a highly demanding world, where customers are extremely picky, and they demand excellent service all the time. But a business that attempts to give them this high quality service telephonically without the right systems is place is foolish. You are costing yourself an enormous amount of money, time and effort.

Every time you make a phone call to a customer using an out-dated system, you are wasting money you do not have to waste. Your staff are spending longer and longer on the phone in an attempt to serve your customers, but it's taking them away from their other work.

Productivity is moving at a snail's pace, because your staff is always on the phone. No matter where you look, there never seem to be enough telephones. You install more and more, at additional cost. Your bills skyrocket. Your productivity plummets. Customers become unsatisfied with your level of service.

Is it because you have not gone out of your way to give them a good customer experience? NO! It is because you have not updated your business telephone system. This system could organize how you communicate with your customers. It will save you money. It will save your employees time and make them more productive because they can be mobile.

There are solutions that exist for these operational problems in the communication niche; it is just that not many business owners know about them. Instead they struggle on, hiring more people, and putting in more phones. You see it a lot at restaurant level. But this 'excessive service' does nothing but damage your reputation and business growth.

When that is damaged your bottom line stagnates. Do not let this happen to your business. You must update your business telephone systems to empower your staff to give the correct level of service in a proactive manner that IMPROVES business.

This is how you will be able to grow and open second and third branches of your company. Then you will really need a quality unified communications infrastructure! It begins with the admission that you are in need of a serious upgrade.

INCREASE PRODUCTIVITY

If you need to increase productivity desperately, but no matter how hard you try you cannot seem to get it right, then an organized, updated business telephone system can do that for you. There is no magic bullet when it comes to human productivity. You can only give people the TOOLS they need and train them to use these tools correctly.

But there's an opposite side to this coin. If you do not give your staff the tools they need, or you give them tools that hamper their performance, productivity is going to sink. Thanks to rapid technologies like VoIP, you do not have to worry about finding ways to make your staff more productive.

With the right business telephone system in place, the communication in your business will improve and along with it —productivity. Your staff will be able to do their jobs better, faster and with more accuracy while simultaneously keeping your customers happy.

Productivity improves service offerings across all profiles. There is less maintenance needed by your staff, the new system is more reliable and flexible – plus it allows for mobility and freedom of movement. In an office environment it is crucial to promote a culture of constant improvement – which is done simply these days via your communication strategy.

Constant improvement means measurement, which is exactly what you can do with new telephone system technologies like call accounting and call recording. If you want your sales people to be better at selling, this is how to do it. No book can teach you how to optimize your business—only analytics can.

Luckily, to increase productivity and make it sustainable all you need to do is upgrade your business telephone system. It sounds too easy – but that is the truth. Better communication leads to higher productivity levels from every single staff member.

They are more connected, aware and motivated to keep your customers happy. When communication is simple, productivity sorts itself out. That is the real magic behind a corporate telephone system. If yours is lacking, you will know. If it is not actively improving your business productivity, you will know. Then it is time to upgrade.

IMPROVE CUSTOMER SERVICE

Customers are at the heart of everything you do as a business. You want your customers to love you, and come back to buy more of what you are selling—again and again. If you can figure out how to do this, you have yourself a winning business model.

But customer service improvement is impossible if you can't even talk to your customers in an efficient and measurable manner. That is where a quality business telephone system comes in handy. It allows you to reach out to customers, and for them to reach out to you.

If your business telephone system is not improving your level of customer service, then it is not doing its job. A customer should be able to call you anytime, from anywhere and get through to the person they want to talk to. The experience should be informative and pleasant. It should not take longer than a few minutes.

There are several things happening in this sequence all at once—you are keeping your call durations down which saves you money. You are enhancing the customer's experience by using an auto-attendant that routes them to the correct person via a reliable IP network connection.

By the time the customer gets to the client service agent, they are greeted by name and given a superb quality of service. That customer buys from you again. They all do. That should be your goal for a dedicated business telephone system. It needs to get your customers from A to Z in the least inconvenient manner.

For your staff, it is a simple matter of answering the call—wherever they are. If they cannot do this, then you have an out-dated business telephone system. Instead of what was just described here, your customer calls, gets put on hold, gets transferred, gets asked the same information four times, gets their call dropped and on the fiasco goes.

It is well into the twenty-first century now, and that kind of 90's behavior is not an asset to your company. Even though the old schools of thought support the 'low cost' system models, there are newer models that can help your small or medium-sized business afford a better quality business telephone system.

NEW "MUST HAVE" FEATURES REQUIRED

Another sign that your business telephone systems need to be upgraded, is the increasing need for 'must-have' requirements and features. There is no denying that we are currently living in a technology boom period, where newer models are coming out faster than the older ones can even be used properly first.

As technology evolves several things happen—it finds a way to make operations more affordable, more effective and less cumbersome for the people working alongside it. This is true of business phone systems.

New must-have features like automated attendant for example are becoming the new 'standard.' This is because auto-attendant replaces the out-dated switchboard operator, an employee position that used to cost businesses a fair amount of money.

But technology has found a way to route calls automatically, in a user-friendly manner that does not require this physical position in the company. As a result, the company saves money, rapidly boosts productivity (the auto-attendant is faster than operators are) and ensures that all customers find their way to a call agent in a few short minutes.

This leads to increased customer service, and as a result customer loyalty. The bottom line is always affected by the subtle changes. An unwillingness to upgrade means that you have out-dated system components that are COSTING you money, and keeping your business from blasting off into a whole new league of service and sales.

Features like call accounting and call recording are two more examples. The technology exists, but if you do not have it you cannot consistently improve your telephonic sales based on the analytics data that you get from these programs.

If you do not have 'must-have' features like the ones discussed in this book, then that is a clear sign that your business telephone system needs an upgrade. With progress in technology comes improvement—not for the technology itself, but for the businesses that use this technology to better manage their communication strategies.

Take a closer look at your current business telephone system. Are you lacking features like commercial on hold, find me/follow me, auto-attendant, client pop up profiles and a web portal for call review? If you are then your system is hopelessly out-dated and in serious need of a total revamp or upgrade. Your competitors have already done it.

CONNECTING MOBILE WORKERS

If your business telephone system does not support multiple businesses or mobile users, then it needs to be upgraded. There is only one reason why you are in business, and that's to make money. Everything else is secondary to that goal.

Having a business telephone system that connects your employees with each other, and makes it easy for them to stay in touch when they are away from their desks, or out of the office, is CRUCIAL to enhanced productivity and increased sales.

Your work force needs to be highly mobile so that your front-end sales staff who are always out at meetings, seeing clients or working off-site, know what is going on at the office. They need to stay permanently connected to your other staff members, to ensure that everyone is doing their job. This is especially true for executives and managers.

What happens if you have several branches of your brand, but no unified communications infrastructure? It is a total nightmare keeping everyone in touch and informed when there is not a central network that everyone can use.

A great business telephone system should be able to accept all calls to all branches, and then route them to the appropriate people as needed. Every individual employee should have a phone number, and an extension that makes it easy for customers to find them.

There is nothing more frustrating in this modern day and age to phone a company, only to be put on hold, bounced around to 3 other people, and 20 minutes later still not have reached the person you need to talk to? That should not be happening to businesses that have the correct infrastructure in place.

Connecting mobile workers is pivotal to the future growth of your company. If you do not think it is important, just think about the dozens of competitors that are getting it right—and how they are taking your customers from you. Communication is a powerful tool, but if your system is not working properly it can be very damaging to people in your company.

The only way to remedy this issue is to upgrade your system. Then your employees can have the mobile communication they need to stay effective in a highly demanding work arena. Half the battle is giving them the tools they need to do their jobs properly. You will immediately see a difference in productivity once you've upgraded!

OFFICE RELOCATION OR EXPANSION

You will know if your business telephone system needs to be upgraded because office relocation or expansion will be impossible without significant financial investment. If your business cannot expand and handle the number of incoming and outgoing calls – then you have hit a fork in the road. Your system is out-dated.

Even though network congestion is quite common, the same thing is happening with data transfer in older telephone systems. What happens is that packet latency occurs which makes delivering data a huge frustration. Then voice latency happens, and you can barely hear the person on the other side of the call.

Your business needs to be able to expand and add as many phones as it needs in a simple and affordable manner. If it can't do that, then you need to upgrade or soon you'll suffer the increasing costs of adding lines, phones and staff to try to balance out demand.

Office relocation can always be a mission – which is why VoIP technologies were created and the physical lines and hardware required for a good business telephone system has been greatly reduced. Now it is incredibly easy to relocate to a new office, without the hassle of having to have hundreds of new lines installed, not to mention the thousands of dollars you will be paying for the phones themselves.

If you cannot pick up your office and relocate it without a significant telephone system debacle, then you need to upgrade your systems now. If you do not, you limit how large your business can grow, and perhaps even inadvertently prevent it from growing.

A modern day office should be able to function at any location. The people need to be connected, the customers need to be connected. But just because this is a prerequisite to success doesn't mean that you need hundreds of wires and phones to make it work. There are simpler solutions that have been invented, that you need to integrate into your business model.

An accomplished business telephone system is no trouble to relocate to a new office. It is also the easiest and most scalable thing in the world. Add as many phones or headsets as you like without technicians or expensive hardware costs.

CONNECTING MULTIPLE SITES

When you own several businesses and they all need to communicate with each other, not having an updated system can be a serious problem. If your business telephone system does not effectively connect all of your branches, then it is not doing its job.

There is a reason why businesses run all of their calls through a centralized network – it saves them money. It reduces costly phone bills because many of these systems run on VoIP networks that charge significantly less for local and cell phone calls.

When you transport your voice calls through an IP network— which also allows you to send and receive data—you can save thousands of dollars per branch, per month! Multiple site connection is a nightmare with out-dated business telephone systems. The costs alone on connection and expansion are enough to force an owner to upgrade.

With a unified communications infrastructure all calls will be run through your PBX or IP network that then automatically distributes the call to the appropriate extension and person. VoIP telephony is especially effective for multiple site connections.

Above and beyond the cost savings are the incredible correction of common inefficiencies associated with owning and running multiple sites. You do not need to have a receptionist answering the phones for you, the auto-attendant does it automatically.

You can also route the calls to one receptionist instead of say 5, from all five locations. From there this core receptionist can assist your customer in solving their problem. Call traffic is also managed far better in these upgraded systems. You do not want to have to deal with a loss of productivity and income because of call congestion.

If your sites are not connected by a unified communications infrastructure or a modern IP PBX system, then you are trailing behind. There is no excuse for a poor communications system in these modern times—they are affordable and available to everyone.

Whether you decide to lease the technology to keep it permanently updated, or buy the latest system so that you can guarantee the successful growth of your business – you can bet that in a year your customers will flock to businesses that have had the foresight to do this.

COST SAVINGS

How much does your business telephone system cost you right now? Does it constantly need repair, or new parts or expansion? Then you are not taking advantage of the new business telephone systems out there, that allow you to do these things at no cost.

There are cost savings everywhere you turn when you implement a new business telephone system. The basic savings come from several areas, namely the fact that you are able to repair them, change them or improve them on your own.

Many of these systems are so easy to use that you do not need to hire a technician to maintain them for you – it is as easy as logging in and clicking a button. Then there is the cost savings that you get from eliminating unnecessary employees and high expense technology.

No more switchboard operators or new lines, and no more visits by costly phone company technicians to train, fix, replace and expand your system. The great thing about new business telephone systems is that they also make your staff more productive, which leads to even more cost savings and income increases.

When a sales person can get up and walk around, while still taking calls—they are being more productive for your business. Every call that they take is a satisfied customer that will buy your products or services.

New business telephone systems result in a cycle of improvement, which leads to consistent and sustainable growth. This growth means that they help trim the fat and enhance the great services that you do provide to your customers.

Of course, these are just some areas where you save money. There is no real calculation for the amount that you save in the long term. How do you calculate happy customers? The cost savings will give you the power to grow your business faster, by investing the money in areas that matter.

Your communication strategy needs to save you money, and make you money. That is why we communicate with our clients, as a business. If your business telephone system does not do this then it is COSTING you money. And you can't afford a system that weighs you down.

CHAPTER 14

Telephone System Maintenance Agreements

"The two words 'information' and 'communication' are often used interchangeably, but they signify quite different things. Information is giving out; communication is getting through."

– SYDNEY J HARRIS

A telephone system maintenance agreement is a standard contract that you enter into with your chosen vendor, and it details how they will maintain your system for you. It is important that your business telephone system stay upgraded and in working order.

But it goes beyond simply maintaining your system—it ensures that if your systems ever crash or break, you can get them up and running again in no time. For every minute your system is down, you lose money. That is why your vendor has to be very good.

WHAT IS A TELEPHONE SYSTEM MAINTENANCE AGREEMENT?

A telephone system maintenance agreement is a document that is signed by you and your maintenance provider (usually your chosen systems vendor) – and it is an agreement on the terms and conditions involved with keeping your system healthy.

These contracts are usually signed when a new business telephone system is installed, or an old system is significantly upgraded. They last for the period outlined, but only after the manufacturer warranty has run out. That means that you are covered for system maintenance during your warranty years and the contract years.

Sometimes fees need to be paid to the business telephone systems company so that maintenance is ongoing. If this is the case, the amounts are outlined in the telephone system maintenance agreement.

These agreements are binding and are very important to both the business owner and the company that is providing you these services. They detail things like obligations, restrictions, recertification's, prices and payments, lapsed service fees, taxes, warranty information, general provisions and limitation of liability.

This agreement is usually discussed once a business telephone system has been picked out and explored by the owner and vendor. The vendor will try to make sure that the communications system chosen functions according to the needs of the business.

Telephone system maintenance is when your vendor provides you with a service that helps you reduce organizational costs, minimizes downtime risk and loss of business and allows you to have a single point of contact for your new system.

WHAT TO LOOK FOR IN A GOOD AGREEMENT

An upstanding telephone system maintenance agreement needs to have certain things in it that protect you as a company, or you could face real trouble if the system fails. That is why you need to know what sort of points to look for in any quality agreement.

- *Look for good value* – all good maintenance agreements provide additional value to the client (business) so that they are able to leverage these areas to improve operational control and business processes.

- *Look for continuity of service* – your agreement needs to outline how long it is valid before it needs to be renegotiated or renewed. Contracts that do not clearly state when they run out can cause a lot of headaches for businesses. Rolling twelve-month contracts are usually the best.

- *Look for fixed price* – if you plan to use your new business telephone system for longer than 3-5 years, you can try and setup a fixed price for the contract over this period. It usually works out a lot cheaper if you do it this way.

- *Look at the exit clause* – all agreements have an exit clause. That way if you are tied to the contract, but want to end it for some reason, you have an out. Otherwise you will be paying that maintenance company by law, for the entire contract period.

- *Look at the upgrading ratios* – your agreement should contain details about upgrades and updates, as they are an important part of maintaining a current, working business telephone system.

- *Look at level of service* – What kind of service level will you receive according to the contract? Review when the company will help you after an emergency, how long they will take to arrive, and when you expect the problem to be fixed.

- *Look at sundry components* – the agreement does not cover old parts of old systems, and these often break and cause issues between owners and maintenance companies. Keep in mind if it was there before they installed the new system – they are not responsible for maintaining it – even if it pulls down your entire system.

THE IMPORTANCE OF FINDING THE RIGHT AGREEMENT

Telephone system maintenance agreements come in all shapes and sizes. There is no 'one size fits all' contract, even though these telephone companies want you to accept their standardized contracts. Often these contracts are in their favor, though they are fair to the client as well.

This does not mean you should sign the first contract you come across. Finding the right agreement is a very important part of maintaining a quality business telephone system for your clients. If you want all of the benefits we have been talking about in this book, then you need a company that will pledge to help you keep it all together.

The right agreement will not necessarily be in your favor, but it will certainly protect your business interests. The interests of your business as this point are simple—to keep your new system operating at maximum capacity to allow for consistent improvements in communication across your company.

The right agreement will also perform the following tasks for you, in order to keep your telephone systems up and running:

- The vendor you have chosen is an authorized dealer and has certified techs for the system you have.

- The vendor you have chosen will carry the spare parts you need in case of emergency. That means they can simply grab the part, and drive out to fix the issue. There is no waiting period for an ordered part.

- Ensure that remote testing and monitoring is available for your system at all times. If you have an issue, your vendor can just remote in and have a look. They will fix problems in a flash, because of this.

- Make sure you outline what is and what is not covered under 'maintenance.' An office relocation for example, may not be outlined in your agreement and could cost you extra when you move.

- Find out the payment structures for expansion, and get discounts and deals for secondary locations in writing.

- Make sure the agreement covers the entire system – that includes the main control unity, on hold functions, voice processors, cabling, communication devices and proprietary system handsets.

- Find out what level of cover the vendor is offering. It should be a 24 hour hotline, with emergency response times. Do not accept contracts that do not agree to 365 days a year cover.

MAINTENANCE SERVICE LEVELS

A maintenance service level is the level of service that your vendor will provide you, after you have signed and sealed the contract agreement. Every company is different, so do not assume that you are 100% covered. Always examine the agreement – especially the area where they detail their service levels.

Maintenance service levels deal with the LEVEL of coverage, response times and any customized solutions you may need in addition to the original contract. Never be afraid to add in points of your own.

- *The 24/7 package:* This is very standard in the telephone systems maintenance industry. The contract should state that you have access to maintenance services 24/7, 7 days a week.

- *Response times:* Make sure that your maintenance company responds to your emergency call within 2-3 hours. This needs to be in the agreement, or you could lose entire working days to a technical systems fault.

- *Certain days*: Many vendors will outline when and how they operate. For example, from weekdays they operate between 9-5. On weekends they operate between 10-2, excluding emergencies of course.

- *Customized solutions*: If you have any specific service requirements that you need, like additional on-site support or training—add them to the agreement.

- *Hardware and software failure*: The level of service must extend to fixing hardware issues in person, and software issues remotely as required. A clear understanding needs to be laid out so that everyone's expectations are properly managed.

- *Objectives of service*: Outline what the vendors objectives of service are, so that you can compare their level of service with their objectives as they work, to ensure quality of service.

These maintenance service levels are a vital part of the telephone systems maintenance agreement. Without a clear understanding of their level of service, you could end up fighting a losing battle, when things go awry.

These levels of service must always be measured, and as a business owner it is your job to make sure that your vendors are holding up their side of the agreement. If you let them, many companies will relax their response times. This is unacceptable for you!

WHAT SHOULD AND SHOULDN'T BE COVERED

When looking for the ideal telephone systems maintenance agreement, certain things need to be covered. Then there are a few things that do not or should not be covered, and we will address those here as well.

Coverage in these agreements depends on the business, the way they run their business and the type of service they agree to provide to their customers. No two businesses are the same so make sure that you read your service level agreement at least three times.

What Should Be Covered:

- The KINDS of fault that the maintenance technicians are able to fix. This includes all manner of breakages to hardware and software glitches.

- Minor and major fault response times. Many vendors have different response times to faults depending on their level of severity. Major faults should take no longer than three hours to tend to, minor faults no longer than eight hours.

- Vendors will often 'swap' their new equipment for your old equipment so that it can be refurbished. They will take your faulty equipment and keep it, supplying you with new equipment for free.

- Software upgrades are incredibly important and should be covered fairly extensively in the agreement. Remote programming also needs to be outlined.

- All charges, surcharges and extra expense scenarios need to be detailed in the agreement for the business owner to see in black and white. That way there will never be any hidden or surprise costs.

- The agreement should also cover terms and conditions, exit strategy, obligations, service levels, restrictions, charges, certifications, upgrades, area availability, warranty details, taxes and general provisions.

- The wires in the walls, jacks, and phone cords should also be covered, even if they weren't installed fresh with your phone system.

- The wires in the walls, jacks, and cords for you fax machines, credit card machines, and alarm system, even if they aren't connected to your phone system should be covered, though that might come at an additional expanse.

- Vendor meets with your carrier

What Shouldn't Be Covered:

- Certain KINDS of faults like normal wear and tear and damage from natural disasters like tornados, fire, floods and storms. You will have to insure your equipment for these specific purposes.

- Anything that does not correlate with three other telephone system maintenance agreements from other companies. The easiest way to see what this vendor has added in their favor, is to cross check their agreement with other agreements. Get your lawyer to read over the details and make a list. Then decide whether or not you can live with the extra conditions, or change them!

GETTING THE BEST AGREEMENT FOR YOUR COMPANY

The best telephone system maintenance agreement is a customized solution. You will sit down with your maintenance vendor and discuss what you expect from them in the coming years, in terms of support, training and overall maintenance.

You also need to outline what 'maintenance' means so that you both have a clear understanding of the word and list of included services. Getting the best agreement possible is part of a process of elimination and comparison.

I never read an agreement without having a few comparable agreements from other companies nearby. This gives me an additional perspective on the kind of maintenance business I'm dealing with, and if they will actually fulfill their contract obligations.

Once I have adequately reviewed the agreement and made sure I am happy with all the inclusions and details, I compare it with the other agreements. I make a list of what is different and if it is not clear, I will query it with the maintenance contractor at the next meeting.

I will also mock up a list of inclusions that I want, based on advice from my lawyer. I will handwrite these inclusions into the margins of the original document. Once I am happy with the final version, I will hand it back to the maintenance contractor.

Then he or she will review it, and get back to me on the changes. We will review it together and agree on the additions and alterations. The contract has now been streamlined to suit my needs as a business owner, yet they are still fair to the contractor.

Getting the best agreement for your company has a lot to do with the vendor that you choose. If they are reputable and have a lot of references that you can call (do call them), and they check out, then you will be alright. What you want to hear is that they are great maintenance vendors that go above and beyond for their clients.

There is always middle ground that you can meet with a business-minded vendor. They want you as a long term customer, so do not be afraid to negotiate price either. The longer the contract, the more discount you can request.

It is common practice to offer new clients a small discount for signing a long term contract with a maintenance vendor. Even if they seem reluctant, they do it all the time. If you are serious about it, you can save yourself money.

CHAPTER 15

The Importance of Your Cabling Infrastructure

"Email, instant messaging, and cell phones give us fabulous communication ability, but because we live and work in our own little worlds, that communication is totally disorganized."

\- MARILYN VOS SAVANT

When planning and designing your business telephone systems, your vendor will be charged with structuring your cabling. As part of a unified communications infrastructure, this is a very important part of your system.

If you ever want to make changes or additions to your data center, then you have got to focus on flexible and scalable structured cabling. Having a well-planned, organized cable strategy is crucial during the periods of installation and maintenance.

CABLING INFRASTRUCTURE DEFINED

A cable system is like a giant network of roads that carries data and voice communication to its end users. As you can imagine, the structure of your cabling infrastructure is very important to the end service.

If the 'roads' are full of potholes and bumps the voice and data packets will not have a smooth journey and will degrade along the way. You must ensure that when your vendor installs your new business telephone system, the structured cabling is done to perfection.

They are basic building blocks of a good system—get them wrong, and you will never stop tending to problems. It does not matter how much bigger and more impressive you make the front-end system, your cabling infrastructure will never give you the performance you need if it's not done correctly.

A cabling crew needs to be brought in to have a look at your existing cable infrastructure. If it is too out-dated or incorrectly installed, a plan has to be made. There is simply no point in adding a new business telephone system to a cabling structure that will not work.

Building a network is difficult enough, but there are several different types of networking cable that can be chosen and used in your system. You need to discuss your options with the cabling crew foreman to see if you can build a structure that helps you maintain a quality business telephone system.

Cabling infrastructure is just as important to your system as the hardware or software that you will be using, or the services you will be receiving from your maintenance vendor. Focus on this chapter and learn all you can about building a quality network of cables.

WHY IS PROPER CABLING INFRASTRUCTURE IMPORTANT?

Think of cabling infrastructure as the foundation of everything. Without it, you cannot hope to transport voice or data packets anywhere. You can also think of structured cabling as the circulatory system in the body of your telephone hardware.

The data center is the area that processes, routes and stores data in your network. It is important that you maximize your network efficiency—due to the increasing demands of modern day business applications. Every second, a new technology is being replaced with something better. There are data-intense applications that need to be utilized by this network in order to help you give your clients a better quality of service.

That is part of the reason why cabling infrastructure is so important. It is really the neural network that connects the brain to the body. Features like data center rack space, cable pathways, installation, maintenance and cooling are also becoming increasingly important to businesses.

You need to design your cabling infrastructure with a competitive advantage and lower cost of ownership to make it worthwhile. Higher speeds will be needed in the future for all data centers. If you can design your cable infrastructure to handle the mounting need for faster information delivery and more reliable connections—you will have done well.

Then the opposite is also true. An average or poorly designed cabling infrastructure can destroy your business. It will degrade network performance, losing as much as 5% packet information on transfer. This means you will get dropped calls, missing data from file transfers and hassles that will never end.

Expect rapid changes in the future to networks and structured cabling. This is why it is important that you have the latest designs. Technology is moving at such a blistering speed that you want to be in compliance with the latest tech as it progresses. Otherwise your out-dated, slow networks will eventually kill your business.

For call centers and sales companies that rely on their network capabilities, this is not a joke. The larger your need for reliable, secure and high performance network speeds, the more you should focus on your cabling infrastructure. It all starts here for you.

CHOOSING WHO INSTALLS YOUR CABLING INFRASTRUCTURE

This leads us to the question – who is going to install your cabling infrastructure? First of all, it needs to be someone highly qualified in structured cabling solutions. You should put a fair amount of research into uncovering what the best possible cabling infrastructure would be for your company.

That way you have something to measure against the professional advice that you will be getting from your chosen expert. There is no point in spending money on new equipment, hardware, software and other telephony devices when your cabling system is going to be handled by 'anyone.'

You can hire an expert to design your company's very own network cabling infrastructure, if that is the route you want to take. Otherwise your cable company will be doing the installation for you. Be very careful of these 'cable companies'—even if you've spoken at length with the superintendent, they are often not even on-site when the cabling is done.

This is exactly why so many buildings suffer from poor cabling infrastructures. Many of these companies even subcontract the work to third party companies, that do a tacky job. To make the most profit, the work is outsourced to a barely functional company that does the cabling for them. This is not what you want to have happen.

It is a better idea to choose a local company that has technicians you can speak to and work with about structured cabling. This is how large Fortune 500 companies ensure that all of their network requirements are adequately met.

An IT manager may also be useful if you have one, and they can work with the technicians to design and build your cabling infrastructure. Always check for the right qualifications and for proof of past work in any contractor that you hire. But be mindful that qualifications don't necessarily mean they can do good work.

It is an even better idea to get the telephone systems vendor involved in the process. Then they can work together to ensure that you get maximum system functionality from the networks. This high performance strategy will be exactly what helps your business grow in leaps and bounds over the next few years.

GUIDELINES TO STRUCTURED CABLING INFRASTRUCTURE

Structured cabling infrastructure is just as important as the structure of the other 'main' utilities like water, heating and electricity. Interruptions to this infrastructure can have unpleasant, even costly consequences.

That is why it is important that structured cabling uses high performance components, and offers long-term support and delivery for high volume networks, like the business telephone system you are planning to install.

A buildings generic cabling structure can be broken down into campus distribution, building distribution, floor distribution and telecommunications outlet. The campus backbone cabling subsystem comes first, then the building backbone cabling subsystem and the horizontal cabling subsystem.

- The amount and variation of subsystems in a structured cabling system depends on the size and location of the building. It also depends on the intention or strategy of the user that is installing the system.

- There is usually one campus distribution per location, one building distribution per building and one or more floor distribution areas per floor space.

- The campus backbone system is a cabling system that provides data and telecommunication services between buildings. It connects the building distribution by way of fiber-optic cabling.

- The building backbone cabling system provides telecommunication services between the floors in a building. It connects building distribution with floor distribution and has copper links and fiber-optic cables.

- The horizontal cabling system provides telecommunication services from the floor distribution to the telecommunications outlet. It is always copper cabling, but fiber-optic cables are also sometimes used.

A well-planned structured cabling infrastructure is key to any successful business telephone system. You could say that it is impossible to install an effective system if these cabling solutions are not first checked, fixed or re-installed. Get the honest opinion of your telephone systems vendor, they will know if your cabling needs to be restructured.

In the meantime, take a closer look at the cord structures used in the building design. These cords will be fixed again and again by your maintenance contractors, so it is worthwhile to ensure that they are in a usable condition. It is a good thing that most cables can be mixed and matched, and they will work together to give you the performance you need.

HOW TO IMPROVE THIS INFRASTRUCTURE

There are many ways to improve a structured cabling system. Your first move will be to check with the vendor whether or not the existing cable system in your building is any good. From there, you can take steps to assess what needs to happen, and how much it will cost to have a proper cable infrastructure installed.

- *Label your cords:* Cabling systems can become very complicated, especially when different companies have installed them. A good way to eliminate confusion is to make sure that your vendor labels every cord. That way you will be able to see where they go, and which panels they are connected to.

- *Quality should be a central focus:* Substandard installations for cabling are actually quite common. It can be frustrating for a new company to arrive, only to find out that there is poor installation throughout the entire data center. Cheap cabling companies are not the answer.

- *Manage your cords correctly:* Proper cord management is key to getting maintenance and repair guys in and out in no time at all. Make sure that you keep patch cords organized so that they can be found at a glance. Otherwise the maintenance men will have to spend ages digging in folds of cords to find something specific.

- *Always choose an appropriate design:* Any old design will not do in the case of structured cabling. A well-planned design will help keep things moving forward when issues do occur and the data center does need to be cracked open. If you choose cheap design over quality, you will never stop having endless issues with your cabling infrastructure. It is the very last thing you want to deal with.

Quality cabling infrastructure improves network speed and guarantees that you get the maximum proficiency possible from the system you have chosen. Remember that hardware, software and cabling are the three most important elements of your system.

You can also have a chat with your vendor and see if they have any ideas on how to improve your current or potential cabling infrastructure. Mention that you want to make it easy for maintenance people to fix problems, and that it is very important to get all the network performance you can from the structure.

CHOOSING WHICH CABLING INFRASTRUCTURE TO USE

When it becomes time to choose which cable infrastructure will best suit your needs, a few things need to happen first. You need to do some personal research into cabling infrastructure, and you need to consult your vendor for advice.

Once you have done this, you will be able to make an educated decision on your cabling infrastructure. There are several different network cables that you can choose from:

- Twisted-pair cables
- Fiber-optic cables
- Stacking cables
- Thin coaxial cables / thick coaxial cables
- AUI cables

Existing Installation

Existing installations depend on what cable types have already been installed. If they can provide the correct bandwidth and capacity for your business telephone system, then there is no problem. You can also mix cable types—running older coaxial cables with twisted-pair cables, if you need too. Otherwise, it is best to install cables in the new style.

New Installation

New installations require 100 meter lengths of 5 twisted-pair cables, with fiber-optic cabling for long backbone connections. Stacking cables are better for short backbone connections inside wiring closets.

Twisted-pair and fiber-optic cables allow you to upgrade in a very scalable manner, from an output of 10mbps to 100mbps network. Because telephone maintenance crews are used to working with twisted-pair cables, they can install the system in a similar manner to how they install the actual telephone system.

Troubleshooting a biaxial cabling system is far more complicated than doing the same thing with a twisted-pair network. If there is a lot of electrical noise, you should use fiber-optic cabling instead. But your structured cabling crew should let you know about this.

Twisted-pair cables are also more reliable and easier to work with than coaxial cables. This is because you can unplug a twisted-pair cable and effect only one connection, but if you unplug a coaxial cable you could bring down the entire system.

Keep in mind that as you are planning your ideal cabling infrastructure, you need to keep later expansion at the forefront of your efforts. Installations are costly, but the cables themselves are very cheap.

FOUNDATION CABLING AND NETWORKING

All telecommunications cabling and networking is the foundation for your business telephone system. There are many subsystems that work together to form a campus system. These data centers are usually run under raised flooring or in special areas in the building.

- *The entrance facility*: This is the point where the exterior plant cables and hardware are brought into the building. These facilities are used for public network services and private network services. This is where carrier and customer meet.

- *The equipment room*: This is where the telecommunications equipment is kept, like the computer equipment, PBX, and video switches. In other words this is where the hardware is kept that serves users on the building premises.

- *Backbone cabling*: This provides connections between the telecommunications closets and the equipment rooms or entrance facilities. You will find backbone cables, cross connects, terminals, patch cords and jumpers here. Backbones also connect buildings together.

- *Telecommunications closet:* This is where the horizontal distribution cables are all connected to their various terminals. The cables are connected to hardware. Cross connection is used with jumpers and patch cords, that make it flexible and able to extend different services to various telecommunications users.

- *Horizontal cabling*: The physical media used to connect every outlet to a closet. The different types of cabling can be used for this. Every cable type has specific performance limitations, cost and size.

- *Work area components*: This extends the telecommunication outlets end ofthe horizontal cabling system to the station equipment.

Filters, baluns and adapters are used to change various electronic equipment items to the structured cabling system.

This is the basic layout of a structured cabling system. The correct cabling infrastructure is incredibly important at the entry-level phase. If you can get this right, you won't ever have to be concerned about long maintenance delays from cabling problems, or slow speeds from poor cabling performance.

The network topology is also established, as either a bus topology, a star-wired topology or a ring topology. Structured cabling was specifically designed to run anything, anywhere at any time – and it works. When you use it, you do not have to follow a set system, things will be compatible with it.

CHAPTER 16

Making the Right Choice

"Communication is a skill that you can learn. It's like riding a bicycle or typing. If you're willing to work at it, you can rapidly improve the quality of every part of your life."

- BRIAN TRACY

A very large part of getting your communications infrastructure right, is taking the time to understand what your options are. This book has walked you through a lot of what you need to know about business telephone systems.

Now, it is up to you to make the right choice. This means deciding on the right vendor for your company, who the best manufacturer is and whether you want to buy or lease the equipment you need to set up your business.

CHOOSING THE RIGHT VENDOR

Before you have even decided on the right business telephone system, you will have to find a good phone system vendor to help you make those decisions. This is why choosing the best vendor is even more important than choosing the right system.

As you have discovered, phone systems can range from basic to horrendously complex, so you will need an informed, experienced vendor to help you put together a system that matches your exact requirements. This is how you should find a decent vendor.

- *Look for on-site installation and training*: Your phone system vendor should install your chosen system for you in a timeous and professional manner. Once this is done, they need to test it, and then train your staff to use it.

- *Look for years of experience*: Your phone vendor needs to be familiar with setting up your system – so familiar that if there are any concerns or problems, they are spotted immediately.

- *Look for constant support*: Support comes in many forms, but with phone vendors you want telephonic support, training support and continuous updates for your hardware and software as time goes by.

- *Look for references and social proof*: The best telephone system vendors are the ones that are proud of what they do. They post customer reviews online and syndicate references and testimonials on their social pages for you to see.

- *Look for a well-known company*: Companies that are well-known are easier to trust because they tend to do a good job. Smaller companies often get away with poor work, bad service and zero support. You do not want that!

CHOOSING THE BEST MANUFACTURER

Always buy from a name you know. Never trust your business to a manufacturer you've never heard of. There are dozens of small independent manufacturers with limited resourses, few dealers, and products that may look good, but don't last the industry standard 12-15 years. Stick with a name you know.

It's critical that you buy a brand that has lots of dealers in your area. There should be no fewer than 5 dealers within reasonable distance. This ensures you'll always be able to get prompt, competitively priced service for many years. If you buy a system from a no name brand company with only one dealer in your area, you will have no where to turn if you ever have a problem with that dealers responsiveness or pricing.

Service from your business telephone system vendor may be superb, but if your hardware or software has problems, you are going to end up losing money. That is why it is important that you choose the best manufacturer of business telephone systems available.

There are many brands on the market that offer telephony solutions, you need to take a hard look at all of them and decide which one will be the most reliable (and affordable) for your company. Then your equipment will be as good as your service.

- In order to find a manufacturer that suits your needs, you first have to know what those needs are. That means making a list of the functionality that you must have in your business telephone system and cross checking it with available functions on all of the major manufacturer websites.

- Compare pricing of each manufacturer. Sometimes a brand will be excellent, but far too expensive for a small business. You will need to balance what you can spend, with what you need – and see if there is a quality manufacturer that offers both.

- See what sort of relationship your vendor has with your manufacturer. If they have a great relationship, your vendor will always be stocked with parts, updates and extras that will make your life easier.

- See which manufacturers offer the best warranties over time. Some offer one year, others two, or five years. If you can see your business scaling up very quickly, then you will not need a five-year warranty because your phone system will be upgraded by then.

- All manufacturers must have a reputation as being reliable, fast and customer-centric. This will make it easier to have things fixed when they break, or to find replacement parts for emergency situations.

The best manufacturer for you will be the one that suits your specialized needs. Some manufacturers are better at creating components for small businesses than others. You just need to put in some time, and do the research to uncover who they are.

Purchasing vs. Leasing

Once you have settled on your business telephone systems vendor and manufacturer, you will need to decide whether you want to buy your equipment or lease it from these companies. As usual, only you can determine which suits you best.

This purchasing or leasing decision will come up during the consultation phase of choosing your business telephone system. You have to keep in mind that by today's standards, equipment goes out of date very fast. Even though your system will work well and last, it may not be helping your business move forward if you do not upgrade.

- If you want to own the business telephone system so that it becomes an asset of your business, then buying a system could work to your advantage. This is especially true if you plan on selling it a year or two later.

- There are many different types of leasing, and they all serve to get you up and running with the best business telephone system—only you do not have to pay for it outright. You pay the leasing fee until your term is up.

There is an old saying that goes 'buy what appreciates, rent what depreciates.' You can use this piece of advice for business telephone systems if you just do not have the money for an outright purchase. With all the different leasing models and payment plans out there, there is no reason why your business cannot afford to have the best system on the market.

CONCLUSION

We've come to the end and this is everything you need to know about business telephone systems! We hope this book has helped open your eyes to the potential that these systems have in your company. Businesses that get their communications infrastructure right from the beginning grow much faster than businesses who stick to the minimum, cheapest requirements.

Imagine if you could rapidly improve the productivity of your business, while at the same time providing the kind of customer service to your customers that makes good brands great! That is the real secret to successful business.

In today's fast-paced world of telecommunications and digital media, you need to be ahead of the curve if you are going to stand a chance at real, sustainable growth. What many companies do not realize is that once your communication infrastructure is in place, everything improves.

The businesses that ignore unified communications infrastructures will be left behind in the global marketplace. But if you take this opportunity to learn from this book and improve your business using qualified business telephone systems – the world is yours!

Start off with good intentions, find your ideal vendor, manufacturer and system – and get moving. The responsibility lies with you. As the business owner, you will need to make the tough decisions regarding budget, installation, implementation and management.

Your team will rely on you for training and correct usage policy. As you uncover how your systems work, set user-friendly procedures that any new employee can learn, and you will never have a problem integrating people into your business telephone system.

With intelligent decisions at the beginning of your business telephone systems process, you will ensure that your company is off to the best possible start. From there, it is just a matter of providing a consistently high level of customer service, while improving your call procedures on a monthly basis.

If you can do this, then your business will become a true member of the global community, with clients all over the world that buy your products and services. This is the future that you need to be guiding your business towards right now.

Connect and communicate!

GLOSSARY OF TELEPHONE SYSTEM ACRONYMS

BRI: Basic Rate Interface

DDI: Direct Dial-In

DID: Direct Inward Dialing

IP PBX: Internet Protocol Private Branch Exchange

IP Telephony: Internet Protocol Telephony

ISDN: Integrated Services Digital Network

PBX System: Private Branch Exchange

POTs Lines: Plain Old Telephone Service

PRI: Primary Rate Interface

PST: Public Switched Telephone

PSTN: Public Switched Telephone Network

SIP: Session Initiation Protocol

UPS: Uninterrupted Power Supply

Virtual IP PBX: Virtual Internet Protocol Private Branch Exchange

Virtual PBX: Virtual Private Branch Exchange

VoIP PBX Systems: Voice Over Internet Protocol Private Branch Exchange Systems

VoIP System: Voice Over Internet Protocol System

VMS: Voice Mail Service

REFERENCES

Introduction To Business Telephone System, http://www.nolayurp.com/introduction-to-business-telephone-system/

Business Telephone System, Wikipedia, http://en.wikipedia.org/wiki/Business_telephone_system

Chapter 1

Quotes on Communication, http://www.leadershipnow.com/communicationquotes.html

Business Telephone System, http://www.webopedia.com/TERM/B/business_telephone_system.html

Find The Right Small Business Phone System Your Company, http://www.cisco.com/cisco/web/solutions/small_business/resource_center/articles/be_more_productive/small_business_telephone_systems/index.html

What To Look For in A Business Phone System, http://www.esi-estech.com/products/systems/what-to-look-for-in-a-business-phone-system.php

Wolski, Chris, *How To Buy a Business Phone System,* http://smallbusiness.chron.com/buy-business-phone-system-2156.html

Lendino, Jamie, *Choose The Right Phone System For Your Business, http://www.cnet.com/4520-10192_1-6419854-1.html*

Choosing a Telephone System For Your Business, http://www.startupnation.com/buyersguides/114/9216/1/1/phone-system-buyers-guide.htm

Feldman, Steve, *The 7 Most Costly Mistakes Companies Make When Choosing a New Phone System... And How To Avoid Them,* http://www.techcare.com/whitepapers/7_costly_phone_mistakes.pdf

Porter, Andy, *Mistakes To Avoid When Selecting Your Business Telephone System,* http://www.invoip.com/blog/bid/159864/Mistakes-to-Avoid-When-Selecting-Your-Business-Telephone-Solutions

Why Choosing The Right Telephone System is Important, http://www.folkersoncom.com/blog/post/Why-Choosing-The-Right-Telephone-System-Is-Important.aspx

Young, Leslie, *Important Factors To Consider in Selecting Business Telephone System,* http://www.slideshare.net/leslielyyoung/important-factors-to-consider-in-selecting-business-telephone-system

Chapter 2

Key Telephone System, http://www.cornerstonetelephone.com/digitalkey

Key Telephone System (KTS), http://www.businessdictionary.com/definition/key-telephone-

system-KTS.html

Telephone System Types, http://www.tel-comp.com/systems-bottom/telephone-systems/

Different Types of Business Telephone Systems, http://www.quote-catcher.com/Module/Article/ArticleView.aspx?idArticle=264

What is a PBX Phone System?, http://www.3cx.com/PBX/pbx-phone-system.html

Small Business PBX: The Basics, http://www.cisco.com/cisco/web/solutions/small_business/resource_center/articles/be_more_productive/small_business_pbx/index.html

What is VoIP and What Can it Do For Your Business? http://www.cisco.com/en/US/prod/voicesw/

networking_solutions_products_genericcontent0900aecd-804f00ce.html

Valdes, Robert, *How VoIP Works,* http://computer.howstuffworks.com/ip-telephony.htm

VoIP, http://www.webopedia.com/TERM/V/VoIP.html

What is a VoIP Telephone?, http://www.3cx.com/PBX/voip-telephone.html

Cloud Based Phone System, http://www.virtualpbx.com/voip/cloud-based-phone-system.asp

Thomas, Dave, *7 Ways a Cloud Based Business Phone System Will Save You Time and Money,* http://www.business.com/blog/7-ways-a-cloud-based-business-phone-system-will-save-you-time-and-money/

Cloud-Based Phone Systems: What's The Deal?, http://grasshopper.com/blog/2012/05/cloud-based-phone-systems-whats-the-deal/

Chapter 3

Rohn, John, *Thinkexist,* http://thinkexist.com/quotation/take_advantage_of_every_opportunity_to_practice/256306.html

What is VoIP?, http://www.voip-info.org/wiki/view/What+is+VOIP

Voice Over IP, Wikipedia, http://en.wikipedia.org/wiki/Voice_over_IP

What is VoIP (Voice-Over-IP?), Cisco, http://www.cisco.com/en/US/prod/voicesw/networking_solutions_products_genericcontent0900aecd804f00ce.html

Mitchell, Bradley, *VoIP – Voice Over Internet Protocol,* http://compnetworking.about.com/cs/voicefaxoverip/g/bldef_voip.htm

Arora, Rakesh, *Voice Over IP: Protocols and Standards,* http://www.cse.wustl.edu/~jain/cis788-99/ftp/voip_protocols.pdf

How Small Businesses Can Use Voice Over Internet Protocol (VoIP) Internet Technology For Voice Communications, http://www.ontariocanada.com/ontcan/1medt/smallbiz/sb_downloads/ebiz_voiceover_protocol_en.pdf

DeSantis, Matthew, *Understanding Voice Over Internet Protocol,* http://www.us-cert.gov/sites/default/files/publications/understanding_voip.pdf

VoIP Configuration, Cisco, http://www.tywf.com.cn/webmanage/eWebEditor/uploadfile/20070701153201956.pdf

VoIP Security Risks Are Often overlooked, Says Forrester Analyst, http://www.infosecurity-magazine.com/view/17676/voip-security-risks-are-often-overlooked-says-forrester-analyst/

James, Peter, Woodward, Andrew, *Securing VoIP: A Framework To Mitigate or Manage Risks,* http://ro.ecu.edu.au/cgi/viewcontent.cgi?

Chapter 4

Communication Quotes, http://www.brainyquote.com/quotes/topics/topic_communication.html

What is a Cloud Based Phone System?, http://www.cloudbasedphonesystem.com/what-is-a-cloud-based-phone-system/

Cloud Phone System, http://www.shoretelsky.com/products/cloud-phone-system/?ce1=2&utm_expid=71849756/

Miller, Ron, *Cloud-Based Phone System Puts Users in Control,* http://www.citeworld.com/consumerization/21248/ringcentral-lets-employees-manage-their-own-phones

Improving Your Business Operations With a Cloud-Based Hosted Phone System, http://www.broadviewnet.com/uploads/whitepaper/OfficeSuite-Improving-Operations.pdf

Nealon, Keith, *Pros and Cons of Cloud-Based VoIP For SMB's,* http://insights.wired.com/profiles/blogs/the-pros-and-cons-of-cloud-based-voip-for-small-business#axzz2Sg67BKC6

Myers, Diane, *The Benefits of Cloud-Based Telephony,* http://www.infonetics.com/whitepapers/2010-Infonetics-Research-The-Benefits-of-Cloud-Telephony-09010.pdf

A Cloud Phone System For Your Business, http://grasshopper.com/cloud-phone-system/

Cartwright, David, *Your First Steps in Implementing Cloud-Based Telephony,* http://www.cloudpro.co.uk/cloud-essentials/3097/your-first-steps-implementing-cloud-based-telephony

Cloud Based Phone System, http://www.voipmechanic.com/cloud-phone-service.htm

Chapter 5

Communication Quotes, http://www.brainyquote.com/quotes/topics/topic_communication.html

Voicemail Features, http://its.ucsc.edu/telephone/voicemail/voicemail-features.html

Voice Mail, http://www.uscellular.com/services/voice-mail/index.html

Bucki, James, *A Definition of Automated Attendant,* http://operationstech.about.com/od/glossary/g/AutomatedAttendant.htm

Automated Attendants, http://www.utexas.edu/its/auto-attendant/

What is an Auto-Attendant?, http://www.3cx.com/faqs/auto-attendant/

Auto Attendant (Automated Attendant), http://whatis.techtarget.com/definition/auto-attendant-automated-attendant

Auto Attendant Features, http://wiki.snomone.com/index.php?title=Auto_Attendant_Features

Telstra IP Telephony Auto Attendant Feature Guide, https://telstra.com.au/business-enterprise/download/document/business-auto-attendant-feature-guide-r17.pdf

User Guide Business Voice Mail, http://www22.verizon.com/idc/groups/public/documents/adacct/west_bus_vms.pdf

Managing Voice Mail, http://www.archives.nysed.gov/a/records/mr_erecords_vmail.shtml

IP Office 8.0 Implementing Voicemail Pro, http://downloads.avaya.com/css/P8/documents/100153493

Chapter 6

Communication Quotes, http://www.inspirationalquotes.asia/2012/06/communication-quotes.html

Top Five Benefits of Unified Communications, http://www.telephonyworld.com/basics/article/top-five-benefits-of-unified-communications/

Benefit From Unified Communications, http://www.cisco.com/en/US/netsol/ns863/index.html

Bailey, Chadwick Martin, *Six Unified Communications Application Benefits,* http://www.itbusinessedge.com/slideshows/show.aspx?c=89082&slide=2

Unified Communications, http://searchunifiedcommunications.techtarget.com/definition/unified-communications

Fontana, John, *Guide To Unified Communications,* http://www.pcworld.com/article/144643/guide_unified_communications.html

A Decision-Makers Guide To Unified Communications, http://www.accountingweb.co.uk/topic/technology/decision-makers-guide-unified-communications

Unified Communications Pocket Guide, http://www.shoretel.com/resource_center/results?#/?content=88425412

How Unified Communications is Increasing Productivity, http://blog.cyclope-series.com/2009/06/10/how-unified-communications-is-increasing-productivity/

Morisy, Michael, *Business Apps With Unified Communications Improve Productivity Costs,* http://searchunifiedcommunications.techtarget.com/news/1357388/Business-apps-with-unified-communications-improve-productivity-costs

Tworek, Richard, *Unified Communications Productivity Study $13,000/year,* http://qovia.com/2013/03/08/unified-communications-productivity-study/

Chapter 7

Communication Quotes, http://www.inspirationalquotes.asia/2012/06/communication-quotes.html

Features and Benefits, http://www.hansensoftware.com/call-accounting-software-features-benefits.htm

Call Accounting, http://www.telmantec.com/solutions/wincall/call-accounting

Campbell, Susan, *What is Call Accounting?,* http://call-accounting.tmcnet.com/articles/204397-what-call-accounting.htm

Jehle, Christine, *What Is Call Accounting?,* http://info.datel-group.com/blog/bid/236948/What-is-Call-Accounting

All About Call Accounting, http://www.ankertek.com.au/about-call-accounting.php

Call Accounting Systems, http://www.tel-electronics.com/call-accounting-systems.php

Features, https://www.tim4biz.com/products/call-accounting/features.aspx

Products Features, http://www.callaccounting.ws/call_accounting_features.htm

Call Accounting Pro, http://www.commone.com/products/capro.htm

Campbell, Susan, *Webinar Explores Importance of Call Accounting With Carrier Call Matching,* http://call-accounting.tmcnet.com/

articles/218026-webinar-explores-importance-call-accounting-with-carrier-call.htm

Chapter 8

Communication Quotes, http://www.inspirationalquotes.asia/2012/06/communication-quotes.html

Wave Call Recording, http://www.vertical.com/mobility/files/V_WaveCallRecording051112HR.pdf

Call Recording Featured Articles, http://www.tmcnet.com/channels/call-recording/

Call Recording, Scalable To Meet The needs of The Enterprise and The SMB, http://www.callcopy.com/products/call-recording

Radman, T, *What is Call Recording,* http://www.trisys.com/blog/?p=41

Majuda Call Recording Features, http://www.dakotapro.biz/enable/phone-systems/call-recording-features.html

Call Centre Helper Reference Guides, http://www.callcentrehelper.com/wiki/A-Z_of_call_recording_features.html

Multi-User Centralized Call Recording and Monitoring For Nortel TDM Telephones, http://www.algosolutions.com/products/call-recording-systems/multi-user-call-recorder.html

Benefits of Call Recording Software, http://www.versadial.com/benefits-call-recording-software/

Vincent, Dennis, *Use Next-Gen Quality Monitoring Tools To Overcome Current Contact Center Challenges,*

http://www.neccf.org/presentations/Sept_2011_Quality_Monitoring_Tools.pdf

Contact Center Recording and Quality Management, http://www.amcomsoftware.com/solutions/contact_center_solutions/Contact_Center_Recording_and_Quality_Management/

Dispute Resolution, http://www.callcopy.com/solutions/business-goal/dispute-resolution

10 Reasons Why You Should Be Recording Calls,

http://www.conversetelecom.co.uk/blog/blog/2012/august/the-benefits-of-call-recording.html

Chapter 9

Automated Call Distribution (ACD), http://www.nd.gov/itd/services/automated-call-distribution-acd

What is Automatic Call Distribution?, http://www.calltrol.com/index_acd.html

Automatic Call Distributor (ACD), http://searchcrm.techtarget.com/definition/Automatic-Call-Distributor

Call Center, http://searchcrm.techtarget.com/definition/call-center

What is The Call Center / BPO Industry, http://call-center-bpo.jobstreet.com.ph/call-center-industry/call-center-bpo-industry.htm

Call Center, http://www.businessdictionary.com/definition/call-center.html

Introduction To Call Centres, http://www.callcentrehelper.com/wiki/Introduction_to_call_centres.html

Chapter 10

Telephone Quotes, http://www.brainyquote.com/quotes/keywords/telephone.html

Polycom Wireless Solutions: Mobility For Today's Workplace, http://www.polycom.com/content/dam/polycom/www/documents/brochures/wireless-family-br-enus.pdf

Netlink Wireless Telephone System Overview, http://www.convtec.com/productpdf/NetLink_System_Overview.pdf

Wireless Telephony Systems, http://www.uc-wireless.com/wireless-solutions/telephony-systems

Benefits of Headsets, http://www.headsets.com/headsets/guide/right1.html

Optional Equipment, http://www.ravon.net/optional%20equipment.html

Introduction, http://esi-estech.com/systems/intro/system_details?id=84

10 Benefits of Paging, http://www.emma-info.org/Technology/10-benefits-of-paging.html

Find Out About The Benefits of Using Paging, http://www.pageone.co.uk/benefits-of-paging

Commercial On Hold, http://www.earglue.com/commercial-on-hold.html

Chapter 11

Carriers Quotes, http://www.brainyquote.com/quotes/keywords/carriers.html

Gelhard, John, *POTs Vs. PRI,* http://www.carrierbid.com/pots-vs-pri/

POTs, http://www.webopedia.com/TERM/P/POTS.html

Your Guide To Choosing Among Many Telephone Options, http://pickocc.org/publications/telecom/Telephone_Service_Options.pdf

Using SIP Trunking You're your Company's Phone System Can Save You a Bundle and Improve Your Customer Service, http://www.cisco.com/en/US/prod/collateral/voicesw/ps6788/vcallcon/ps11370/what_sip.pdf

Session Initiation Protocol, Wikipedia, http://en.wikipedia.org/wiki/Session_Initiation_Protocol

Schulzrinne, Henning, *The Session Initiation Protocol (SIP),* http://www.ee.oulu.fi/~skidi/teaching/internet_multimedia_2002/sip_long.pdf

Voice, Dynamic PRI, http://www.entelegent.com/uploads/resources/documents/EnTelegent%20PRI%20Dynamic-12292011085812.pdf

Flexible, Reliable Integrated Nationwide Voice and Data Services, http://www.earthlinkbusiness.com/_static/_files/_pdfs/EC-SIP-Trunking-PRI.pdf

Primary Rate Interface (PRI), http://www.vita.virginia.gov/services/default.aspx?id=4598

Primary Rate Interface, http://www.cornerstonetelephone.com/pri

Primary Rate Interface (PRI), http://searchnetworking.techtarget.com/definition/Primary-Rate-Interface

Direct inward Dialing (DID), http://searchcio-midmarket.techtarget.com/definition/Direct-Inward-Dialing

What is DID – Direct Inward Dialing?, http://www.3cx.com/faqs/did/

Guide To VoIP Phone Service Providers, Who Can You Trust?, http://www.business-voip-phone-service.com/voip-phone-service-providers/

Young, Leslie, *Benefits of Reliable VoIP Business Phone Service Providers,* http://www.slideshare.net/leslielyyoung/benefits-of-reliable-vo-ip-business-phone-service-providers

Choosing a Telephone Service Provider, http://www.buyerzone.com/telecom-services/business-phone-service/bg-telephone-service-provider/

Chapter 12

Communication Quotes, http://thinkexist.com/quotes/with/keyword/communication/

Housley, Sharon, *Online Shopping's Pros and Cons,* http://www.translationdirectory.com/articles/article2329.php

Buying Online Has its Pros and Cons, http://www.epinions.com/content_1972609156?sb=1

Used Phone Systems, http://www.blueskycomms.com.au/used-phone-systems/

Hatton, Jonathan, *Tips For Rating Office Phone Systems,* http://www.slideshare.net/JonNWO/tips-for-rating-office-phone-systems

Refurbished, http://www.best4systems.co.uk/refurbished-products

New and Refurbished Telephone Systems For Your Office, http://www.usedphonesonline.com/

Pre-Owned Telephone Systems, http://www.quadisvoice.com/Pre-Owned-Phone-Systems.html

Improving Assets and Productivity Using Second Hand Phone Systems, http://business.goodarticle.net/2012/08/17/improving-assets-and-productivity-using-second-hand-phone-systems/

Chapter 13

Quotes About Phone, http://www.goodreads.com/quotes/tag/phone

Duray, Matt, *Tech Communication: Obsolete Telecommunications Systems Can Cost You Money, Productivity,* http://billingsgazette.com/business/columns/tech-connection-obsolete-telecommunications-systems-can-cost-you-money-productivity/article_631ec783-4c12-5eec-bfeb-537adac83928.html

Matt, *6 Reasons You Need To Upgrade Your Business Phone System,* http://www.telcorebates.com/2011/11/6-reasons-you-need-to-upgrade-your-business-phone-system/

VoIP Can Help To Improve Business Productivity, http://www.voip-info.org/wiki/view/VoIP+can+help+to+improve+business+productivity!

3 Ways To Increase Business Productivity With Hosted VoIP, https://www.doublehorn.com/blog/3-ways-increase-business-productivity-hosted-voip

VoIP Increases Productivity, http://www.businessvoipnews.com/?p=70

6 Signs That You Need To Upgrade Your Business Phone System, http://www.windstreambusiness.com/media/1145/upgrade_business_phone_system.pdf

Time To Upgrade Your Business Phone System – 6 Signs That Point To Yes, http://www.sterling.net/blog/time-to-upgrade-your-business-phone-system-6-signs-that-point-to-yes/

Chapter 14

Communication Quotes, http://thinkexist.com/quotes/with/keyword/communication/2.html

Why Your Business Needs Telephone System Maintenance, http://www.dataphone.co.uk/why-your-business-needs-telephone-system-maintenance/

What To Look For In a Telephone Maintenance Contract, http://news.lb-telecom.co.uk/post/41865152959/what-to-look-for-in-a-telephone-maintenance-contract

What To Look For When Deciding on a Phone System Maintenance Provider, http://www.telcom-data.com/articles/159.aspx

Telephone System Maintenance, http://www.phones4business.com/telephone-maintenance/

Asterisk Maintenance and Service Level Agreements, http://www.stuartelvish.com/service_level_agreements.html

System Maintenance Agreements, http://www.myipsystem.com/Maintenance_agreement_business_telephone_systems.htm

Telephone System Maintenance Contracts, http://www.astrocommunications.com.au/telephone-system-maintenance.php

F5 Maintenance Terms and Conditions, http://www.f5.com/pdf/customer-support/maintenance-agreement.pdf

Telephone System Maintenance Information, http://www.telephone-engineer.com/downloads/maintenance-agreement.pdf

Chapter 15

Communication Quotes, http://www.brainyquote.com/quotes/topics/topic_communication.html

Troubleshooting Cabling Problems, http://www.processor.com/editorial/article.asp?article=articles%2Fp3215%2F15b15%2F15b15.asp

Donovan, James, *Guess What? Infrastructure is Important!,* http://commscopeblogs.com/2011/01/27/guess-what-infrastructure-is-important/

Kozischek, David, *The Importance of Cabling Infrastructure Design in The Data Centre of Tomorrow,* http://www.cxo.eu.com/article/The-importance-of-cabling-infrastructure-design-in-the-data-centre-of-tomorrow/

Cabling and Networking Infrastructure The Foundation is Everything, http://www.interweavetech.net/tech-news/2012/10/cabling-and-networking-infrastructure-the-foundation-is-everything/

Choosing The Cabling Infrastructure, ftp://ftp.hp.com/pub/networking/software/c1.pdf

Structured Cabling, Anixter, http://www.scom.uminho.pt/uploads/Apoio%20-%20Doc%20Tec%20-%20anixter.pdf

Information Communications Technology Structured Cabling Infrastructure Guidelines, http://www.ed.ac.uk/polopoly_fs/1.58618!/fileManager/ICT%20Structured%20Cabling%20Guidelines%20.pdf

Cabling Infrastructure Standards, http://www.clpccd.org/facilities/documents/CLPCCDCablingStandards-062606_000.pdf

P, Rick, *Cabling Infrastructure is Becoming Increasingly Important,* http://www.cxtec.com/blog/cabling-infrastructure-is-becoming-increasingly-important/

6 Must-Do's When Installing Your Cabling Infrastructure, http://www.tcsicanada.ca/1/post/2013/02/6-must-dos-when-installing-your-cabling-infrastructure.html

Who Is Installing Your Cabling Infrastructure, http://deployedtech.com/2011/06/who-is-installing-your-cabling-infrastructure/

Chapter 16

Communication Quotes, http://www.brainyquote.com/quotes/topics/topic_communication.html

Choosing a Phone System Vendor, http://www.resourcenation.com/buyers-guides/choosing-phone-system-vendor/

Should I Lease or Should I Buy?, http://www.telephoneguru.net/finance.html

Lease Analysis, Leasing Strategies, Lease Buyout an End of Lease Negotiation, http://www.morganbirge.com/phone-services/buying/Leasing-Buying.php

Leasing Versus Buying, http://myteltek.com/blog/2010/03/leasing-versus-buying/

Made in the USA
Monee, IL
04 November 2026

67386603R10098